CALEB SCHARF

The Copernicus Complex

*The Quest for Our Cosmic
(In)Significance*

PENGUIN BOOKS

PENGUIN BOOKS

UK | USA | Canada | Ireland | Australia
India | New Zealand | South Africa

Penguin Books is part of the Penguin Random House group of companies
whose addresses can be found at global.penguinrandomhouse.com.

First published in the United States of America by Scientific American/Farrar,
Straus and Giroux 2014
First published in Great Britain by Allen Lane 2014
Published in Penguin Books 2015
003

Text copyright © Caleb Scharf, 2014

An excerpt from *The Copernicus Complex* originally appeared, in slightly different
form, in *Scientific American*.

The moral right of the author has been asserted

Printed in Great Britain by Clays Ltd, St Ives plc

A CIP catalogue record for this book is available from the British Library

ISBN: 978-0-141-97493-4

www.greenpenguin.co.uk

MIX
Paper from
responsible sources
FSC® C018179

Penguin Random House is committed to a
sustainable future for our business, our readers
and our planet. This book is made from Forest
Stewardship Council® certified paper.

CONTENTS

THE
COPERNICUS
COMPLEX

PROLOGUE:
FROM MICROCOSM TO COSMOS

It all begins with a single drop of water.

With one eye scrunched shut, the drapery tradesman and budding scientist Antony van Leeuwenhoek stares intently through the tiny lens he has fashioned from a piece of soda-lime glass. On the other side of this shiny bead is a quivering sample of lake water, scooped up during an outing the previous day around the city of Delft in the Netherlands. As he adjusts the instrument and lets his eye relax and focus, van Leeuwenhoek suddenly finds himself falling headlong into a new world, a swarming metropolis of alien design.

Within the previously invisible universe of this single speck of water are arrays of beautifully coiled spirals, animated blobs, and bell-shaped creatures with skinny tails, wiggling, gyrating, and swimming busily with absolutely no regard for his right to be peering in at them. It's a shocking vision: van Leeuwenhoek is not just a human, he is a cosmically huge giant observing another world contained within his own. And if this one drop can be home to its own universe, then what about another, and another, and all the drops of water on Earth?

The year is 1674, a time sandwiched between some of the most profound changes in Western science and thought. A little more than a century earlier the Polish scientist and polymath Nicolaus Copernicus had published *De revolutionibus orbium coelestium*—"On the revolutions of the celestial spheres." In this book Copernicus had put forth his completed heliocentric model of the universe, shifting the Earth from the center of the cosmos to a secondary place, spinning and orbiting around the Sun—a demotion that would reshape our species' scientific history.

In the intervening decades the Italian Galileo Galilei had built his telescopes and seen the moons of Jupiter and phases of Venus, convincing him that Copernicus was right—a heretical view at the time, one that cost him dearly when it attracted the scrutiny of the Roman Inquisition. His contemporary, the German Johannes Kepler, went even further by stating that the orbits of the planets, including the Earth, traced not perfect circles but rather eccentric ellipses, unsettling any conception of a rational universe. And in a little over ten years' time from when we find van Leeuwenhoek gazing through his lens, the great English scientist Isaac Newton will publish his monumental *Principia*, laying out the laws of gravitation and mechanics that will, unwittingly, make the arrangement of our solar system and of the universe at large a thing of austere beauty, untended by any guiding hand but physics and mathematics. It is by any standards an extraordinary time in human history.

Antony van Leeuwenhoek was born into this rapidly transforming world in 1632 in Delft. His early life was relatively ordinary. He never received much education beyond the basics. As a young man he quickly established himself as a tradesman dealing successfully in linens and woolens. He was also a relentlessly interested and curious person, once describing himself as "craving after knowledge," a characteristic that would result in a voluminous legacy of observations and writings about his greatest passion, the microcosm.

Sometime in the year 1665 van Leeuwenhoek came across the great work *Micrographia*, by the English scientist Robert Hooke. *Micrographia* was a phenomenon: the first major publication of the fledgling Royal Society in England, the first best-selling science book, and a cornucopia of the most fabulously detailed illustrations of the magnified textures of everything from insects to minerals, bird feathers, and plant life. It was an atlas of the world seen through a new set of eyes, those of the microscope.

This novel technical art of magnifying objects using a series of lenses had begun not long before, in the late 1500s. The compound microscope enabled the sharp-eyed and sharp-minded Hooke to make his beautiful drawings of all these incredible things that were sitting right under everyone's nose. But even Hooke's best microscopes achieved magnification factors of only ten times to perhaps fifty times. What might be lying even deeper beneath? For van Leeuwenhoek the mystery was impossible to resist, and so he took it upon himself to learn to build the optics necessary to catch his own glimpse of this unexplored realm.

Exactly how van Leeuwenhoek made his microscopes remains a little unclear to this day. He was incredibly secretive and a bit dramatic about it all, beavering away behind closed doors at his home. But from instruments he bequeathed to the Royal Society, and from the accounts of people who visited, we do know that his principal trick was to fashion tiny, perfect beads of glass—probably by pulling molten glass fibers and fusing their ends together. Then he mounted these spherical lenses, with focal lengths of barely a couple of millimeters,

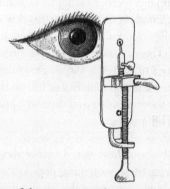

Figure 1: An illustration of the van Leeuwenhoek microscope. Samples could be placed on the tip of an adjustable metal probe just in front of the opening in a plate that holds the glass lens. Bringing it up to the eye completes the optical system.

in small brass plates with screwlike arms that would position a sample right by the lens. By holding the plate across his eye, van Leeuwenhoek could gain some astonishing magnifications, possibly as high as *five hundred* times in the very best cases.

He also didn't just make a single microscope, or even a few. In a remarkably modern burst of innovation, he made well over two hundred. In fact, it appears he made a microscope for pretty much every subject he wanted to study—a customized job each time. And thus it was a few years later, on a September day in 1674, that the tradesman could be found putting a fateful drop of water in front of a lens in its purpose-built viewing platform.

Van Leeuwenhoek's innate gift for fashioning optics took him not to outer space but to a microscopic cosmos, on what was perhaps an equally shocking journey. Within these drops of water he discovered unknown types of living organisms, hidden away from prying humans by simply being too small to see with the naked eye. He also quickly realized that if these minute life-forms could be in a drop of lake water, they could be anywhere, and he extended his investigations to other realms.

These included the fascinating, but rarely appreciated, nooks and crannies of the human mouth and the sticky mix of saliva and plaque gumming up our teeth. Putting these samples under his lens, van Leeuwenhoek found even more diversity: dozens, hundreds, thousands of even smaller "animalcules" swimming about in their rather repulsive oceans. These varied and active organisms offered the first human glimpses of bacteria, the single-celled living things that we today know represent the majority of life on the planet, outdoing everything else by sheer number and diversity, just as they have done for the past 3 to 4 billion years.

I've often thought about how van Leeuwenhoek might have felt when he came across these swarming populations of "animalcules." There is little doubt he was amazed—his notes and writings convey a gleeful pleasure in being able to unveil what was previously invisible to us all—and he spent subsequent years examining and recording more and more specimens and samples. But did he ever wonder if one of those swimming, spinning little creatures was looking back at him? Did he wonder if the occupants of a drop of water were busy asking whether they were the center of the universe, trying to deduce the mechanics of their own heavens, which might have included his great eye hovering above them?

There is no good evidence to suggest that van Leeuwenhoek thought about these questions. People were certainly excited about discoveries like these. But there's not much to indicate that van Leeuwenhoek, or anyone else at the time, stepped back to reflect on any cosmic meaning. It's practically inconceivable to me that no one ran through the streets shouting the news: "We're not alone! We're full of tiny creatures!" But it doesn't seem that people felt their place in the universe undergo a seismic shift with the discovery of these microscopic underpinnings—even though they revealed a layer of reality that didn't include us.

To be fair, this was in part because we just didn't yet appreciate the true relationship between microbial life and our own. It would be another two hundred years, till the mid-1800s, before the idea that bacteria could cause disease was formally recognized. In turn, it would be another century after that before we would appreciate how these denizens of the microcosm are part of our own fundamental composition, swarming within our guts in their hundreds of trillions, intimately connected to our physiological well-being. And even now, in the twenty-first century, we are only just beginning to understand this remarkable symbiosis.

In the 1600s the vast underworld of van Leeuwenhoek's animalcules was accepted as an interesting fact, but one that was largely irrelevant to our own cosmic importance. This narrow viewpoint was not just a product of the times. It was a reflection of a tendency so deeply rooted in the strange and powerful human psyche that it must relate to our most fundamental evolutionary history and our instincts for survival. It's a type of behavior that we all carry with us today—a tendency to automatically assume our significance above all other things, regardless of what evidence is placed right in front of us.

Cultures vary, for sure, with differing degrees of respect for our natural surroundings and our worldly cohabitants, but most of us assume our overall importance more than our insignificance. This solipsistic behavior crops up again and again, despite our endless desire to know how and why we exist. Perhaps we sense that these questions open the door to scenarios that leave us among the filthy and irrelevant chaff of the passage of cosmic time. The most critical example is that of the Copernican Principle, which states that the Sun, not the Earth, occupies the center of the heavens, and that a spinning Earth, as well as the other planets, circles around this fiery orb. It is a worldview that asserts that we are not the center of all existence; we are not "special." In fact, we are as dull as they come.

Indeed, the last five hundred years of science have seen the chalice of our significance shaken more than at any other period in recorded human history. The overlapping revolutions of modern optics, astronomy, biology, chemistry, and physics reveal that we inhabit merely one sliver of nature; that our normal awareness of the world resides in neither the microscopic nor the cosmic, but in what might be considered the narrow borderlands in between. And now, in the twenty-first century, we stand on the cusp of the ultimate in disruptive events: the real possibility of discovering whether life exists elsewhere, beyond the confines of planet Earth. We could find out that we are, after all, just like the animalcules in a drop of Delft lake water—one occupied world among billions. Or that we are as good as alone in the cosmos, a tiny swarm in one crevice of an incomprehensibly enormous mouth of expanding space and time.

Most surprising, we now have some reason to believe that these possible outcomes may also be linked to an even deeper question: whether or not this universe is itself just one instance of a near-infinite array of universe-like entities emerging from the most fundamental characteristics of the vacuum. Some of these ideas are positively head-spinning—inducing precisely the same kind of vertiginous feeling that van Leeuwenhoek must have had when he gazed at the microscopic cosmos for the first time.

Much of this book is about *how* we may get to answer these questions; how our quest to understand our cosmic significance is making practical and tangible progress and, in the process, challenging so many preconceptions and conceits. I will argue that we can already draw some conclusions, and I will present a proposal for how to take our knowledge about life in the cosmos far beyond its present state, to a new level of insight.

To get to the crux of the problem requires a careful dissection of one of the greatest principles to ever serve science and philosophy. The roots of this idea are modest; they are in nothing more than our daily and nightly experience of the sky above.

We'll see how the decentralized reality that Copernicus proposed was logically compelling because it helped explain the detailed motions of the Sun, Moon, and planets across the heavens. And it accomplished this explanation in a more direct and elegant way than preceding theories. But for many people in his time, it was a horrible concept. As well as being theologically unappealing because it suggested we were unimportant, parts of the idea were also scientifically distasteful: they represented a challenge to the very core of prevailing analytical thought about the mechanics of the cosmos.

Over time, we have taken this decentralization even further, and we now consider any scientific theory that depends on a special origin or a unique viewpoint to be inherently flawed. This is eminently sensible. If such generalization weren't true, the laws of physics that apply to you might not apply to your friend who happens to live on the bad side of town, a possibility that runs counter to everything we know. However, as I'll argue, the Copernican Principle may have reached the end of its usefulness as an all-encompassing guide to certain scientific questions.

Indeed, while we cannot be at the center of what we now know to be a centerless universe, we appear to occupy a very interesting place within it—in time, space, and scale. Various arguments have certainly been made along these lines before, sometimes culminating in the hypothesis that Earth is exceptionally "rare," especially in regard to the development of technologically intelligent life. This conclusion is extreme, however—and I don't believe it's been convincingly substantiated. I'll show you why.

Nonetheless, the specifics of our circumstances—our place

between the microscopic and the cosmic, on a rocky planet around a star of a certain age—do most certainly affect the way we make inferences about nature, and the way we search for other life in the universe. The specifics of our own cosmic "address" also provide vital clues. Taking it further, I'm going to argue that for us to make genuine scientific progress in determining our cosmic status, we must find a way to see past our own mediocrity. I'm going to present a way to do this.

The quest to find our cosmic significance, to resolve the conflict between our Copernican mediocrity and our specialness, will take us from the deepest history of the Earth to its farthest future, to planetary systems across our galaxy, and from the great universe of astronomy to the microscopic universe of biology. It's also going to take us to the cutting edge of scientific inquiry into our cosmic origins—an exploration being carried out through mathematical wizardry and cunning observations of nature. And it will lead us to an unwavering examination of the specific circumstances we find ourselves in, our place in the cosmos.

1

THE COPERNICUS COMPLEX

At a rather pleasant spot in the Aegean Sea in the third century B.C., on the vine-rich island of Samos off the western coast of what is now Turkey, the Greek philosopher Aristarchus had just had a brilliant idea. He proposed that the Earth spun and moved around the Sun, placing this scorching solar orb at the center of the heavens. It was, to say the least, a bold notion—Aristarchus's idea of "heliocentrism" was as outrageous in his time as Copernicus's revival would be in the distant future.

Records of Aristarchus's works are fragmentary, and most concern the clever geometrical analyses that he used to argue that the Sun is significantly larger than the Earth. But it's clear that from that insight he arrived at the idea that the Sun was central to the known cosmos, and that the stars were extraordinarily distant. This was a huge conceptual leap to ask of people. It also required understanding a phenomenon called *parallax*.

Parallax is earthbound as well as celestial, and is an easy concept to grasp. Close an eye and hold one hand up, fingers spread and viewed on edge. If you move your head side to side you will see different fingers appear and disappear from behind each other as your vantage point, or your angle of view, changes. This is all that parallax is: the apparent change in where distant objects appear relative to each other, depending on a line of sight. The farther away those objects are, the smaller that apparent change—the smaller the perceived angular displacement between them.

Part of Aristarchus's bold argument involved the fact that the stars in the night sky *didn't* appear to have any parallax; they never moved among themselves. So if the Earth were *not* stationary at the center of all existence, he reasoned, the stars must be so distant, so enormously far away from us, that we couldn't measure their parallax as the Earth moved its position.

Not long before Aristarchus made his ideas known, the great philosopher Aristotle had already dismissed the possibility that the stars were any more distant than planets by appealing to this same lack of parallax, among other things. Aristotle's argument was founded in reason and common sense. It built on even earlier ideas that the Earth was central to existence. The way he put it was simple: if no parallax could be seen in the stars—they did not shift around relative to one another at all—they must be all affixed to some layer of the sky that surrounded us at the unmoving origin.

All of which sounds logical, except that Aristotle's own preferred cosmology (elaborating on ideas from his mentor, Plato) consisted of approximately fifty-five thick, crystalline, transparent spheres concentrically nested about the stationary Earth and carrying the planets and stars about their business. In this geocentric universe we were at the focus of all natural motions, with the stars and planets simply following perpetual circular paths around us as the crystalline spheres slid and rotated.

You might well ask why it took fifty-five spherical crystalline layers for Aristotle to build his cosmology. Part of the reason is that he had to justify a system of cosmic mechanics, a transfer of forces whereby one shell would rub on another, pushing it around—a great scheme of motions and machinery to keep everything tracking through the sky. This structure was intended to deal with the other most critical issue facing would-be cosmologists of the time; unlike the stars, the planets *do* move around the skies in a complicated fashion.

These tricky motions were a major piece of the puzzle that Aristarchus, and later Copernicus, tried solving by displacing the Earth. The word "planet" is derived from the Greek phrase for "wandering star," and our brightly reflective planets most certainly do wander. Not only do they appear to move relative to the stars, noticeably shifting in position as the nights go by; sometimes they reverse course, performing a celestial loop-the-loop over a few months before carrying on. Planets like Venus and Mercury are even more subversive; often they're nowhere to be seen. And even the speed of planetary paths across the heavens seems to be slower and faster at different times, with the brightness of these miscreants changing as well.

So you might think that when Aristarchus proposed his heliocentric system there would have been a huge sigh of relief, because placing the Earth on its own circular path around the Sun quickly provided a solution for much of the curious backward motion of the planets—what would later be known as "retrograde" movements. In this configuration the simple reason for such odd behavior was that our own vantage point was shifting as the Earth itself moved in a circle. There would naturally be times when our motion relative to a planet was either forward or backward, and our distance from a planet would change—lowering or raising its apparent brightness.

It was a lovely, elegant, and fact-based idea—and many people hated it. If the Earth moved, there should be a noticeable parallax among the stars, which surely couldn't be *that* far away. And apart from this lack of observable parallax, displacing the Earth from its vaunted central position was anathema; it was ludicrous to consider that the very hub of our existence was not at the core of everything, and so poor Aristarchus got it in the neck.

The other part of this antipathy toward heliocentricity likely came from distaste for ideas that hinted of pluralism. In opposition to the likes of Plato and Aristotle, who argued for a divine and unique creation of the Earth, Greek thinkers such as Democritus and Epicurus instead advocated a picture of reality rooted in the notion of indivisible pieces and empty void—atoms and space. These atoms weren't atoms as we now know them, but a philosophical concept of units of matter—too small to be seen, solid and uniform within, varied in size, shape, and weight—that could be used to describe an infinite number of structures. The idea of atoms led these thinkers to consider that the Earth was unlikely to be unique. Far from it—there should be an infinite number of inhabited worlds located within an abstract form of space and time and what, in retrospect, amounted to parallel universes. Not surprisingly, the plurality of worlds did not sit well with anyone following the Platonic or Aristotelian schools of thought.

What happened instead was that a number of natural philosophers in the decades following Aristarchus came up with a geocentric "fix" to account for the annoyingly unconventional motion of the planets across the skies, and to keep Earth rooted as the unique center of existence. Their solution to the dilemma of celestial movements probably first originated almost a century after Aristarchus and Aristotle butted heads, with the astronomer and geometer Apollonius of Perga around the turn of the second century B.C.

Later on, this explanation was subsumed into the works of Claudius Ptolemy. Living some three hundred years after Aristarchus, the Greek-Roman citizen Ptolemy resided in Egypt under the rule of the Roman Empire. He was a prolific thinker, producing significant works on many topics, including astronomy, geography, astrology, and optics. And most important, he produced an astronomical treatise known as the *Almagest* that laid out a cosmological vision that would stick for the next 1,400 years.

In Ptolemy's system, the Earth is firmly stationed at the center of the universe. Moving outward are the Moon, Mercury, Venus, then the Sun, and then Mars, Jupiter, Saturn, and the fixed tapestry of stars—all following circular paths. To translate this arrangement into the messy movements seen in our skies, he added a clever set of extra motions along special circular paths called deferents and epicycles. And these were, rather ironically, centered on a location *offset* from the Earth (a peculiarity that seems to have escaped the scrutiny of zealous geocentrists across the centuries).

In this ingenious arrangement, the planets and the Sun move around the smaller perfect circles of the epicycles, which in turn move along the larger circles of the deferents, which rotate around a point separated from the Earth. The end result matches up with the major features of the looping, back-and-forth pathways of the

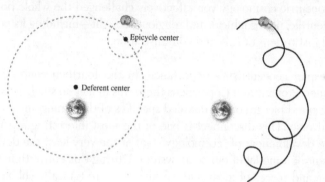

Figure 2: A sketch of one of the simpler versions of Ptolemy's geometric explanation for planetary motions in a geocentric cosmology. Here Mars follows a circular path around a small epicycle that in turn moves around a larger circular deferent. The result? Mars appears to loop back and forth across the sky, getting nearer and farther as it goes.

planets. To do this, Ptolemy's system had to be very fine-tuned to actual observations of the planets. Each and every deferent and epicycle was meticulously sized and located in order to give the best fit possible to the real meanderings of the known worlds.

Even with such fine engineering, the system couldn't quite get everything right—there were little deviations here and there from astronomers' measurements over the years. Planets would arrive a bit early or late to certain positions on the sky—not enough, though, to discourage everyone. Here was a plausible model for the nature and motion of the Sun, Moon, and planets that was geocentric, grounded in the precision and truth of geometry, and in agreement with the thinking of the Great Philosophers. The model comforted mathematicians and theologians alike.

Later, as Ptolemy's ideas were subsumed and integrated into the religious and philosophical doctrines of the Western world in the Middle Ages, they became intricately attached to a unified conceptual framework. Like arterial conduits helping to keep the blood flowing, the geocentric spheres and their epicycles were a key part of the machinery of the perceived universe. If you challenged geocentric cosmology you effectively challenged the whole body of scientific, philosophical, and religious thought—including its powerful institutions of rule and administration.

Despite geocentrism's importance, in the fourteen centuries between Ptolemy and Copernicus there was in fact no single generally accepted picture of the detailed specifics of the arrangement of the universe. This disconnect is one of the most interesting aspects of the development of "cosmology"—or at the very least the development of a model of our solar system. During this entire time span, bits and pieces of ideas and worldviews were typically cobbled together for convenience, as and when needed—a cosmic mix-and-match. It depended on whether you wanted a mathematically driven universe, or a more abstract philosophical one. And all of these ideas in turn reached back to the varied hypotheses and proposals of a multitude of long-departed Greek thinkers.

Equally important for this cosmological history was that so much of its character hinged on the available precision of measurement. Aristotle and Aristarchus were no slouches when it came to making careful astronomical observations, but they were severely limited with only human eyes and basic tools for assessing angles and distances. This limitation meant that they actually had no idea what something like the true parallax motion of the stars really was; they just assumed it was zero.

The data on the motion of the planets themselves was also of limited precision, and it left gaps in knowledge that would let Aristotle and Ptolemy squeeze geocentric models, with their increasingly elaborate geometric arrangements, into the picture. The models may not have been perfect, but humanity's observations of the heavens weren't good enough to disprove them.

So by the late 1400s there had been little real progress in formulating a better model for the motions of the Earth, the planets, and the stars—especially given the accepted need to be consistent with the religious and philosophical doctrines of the Western world. In fact, I think it's fair to say that to our modern scientific eyes, medieval cosmological models were in a thoroughly messy and inconsistent state. The time was certainly ripe for some drastic improvements to be made. All that was needed was the right person.

Nicolaus Copernicus was born on February 19, 1473. Growing up in a part of Prussia that had recently been ceded to Poland, Copernicus had the good fortune to be part of a sophisticated and well-off family. He got an excellent education that included a thorough grounding in philosophy (which by default was the intensive study of the works of the ancient Greeks), mathematics, and the natural sciences—including astronomy. He was also genuinely voracious in his appetite for knowledge, and doesn't seem to have shied away from hard work during his entire lifetime, even producing manuscripts on poetry and politics in addition to his scientific investigations.

His early schooling led him on to further studies in Italy, where he began to get more and more interested in astronomical observations, especially those that related to the measurable deviations of lunar and planetary behavior from the Ptolemaic system. Other investigators of the time were also well aware of these deviations, but the industrious Copernicus was particularly moved to step outside the usual bounds in looking for answers, and was eager to find a more accurate solution than the one that Ptolemy had devised so long before.

In the early 1500s Copernicus drafted what would later become the basis for his full heliocentric model of the solar system—a forty-page work known as his *Commentariolus*, or "little commentary." It was never officially published during his lifetime, but instead a few copies circulated in a limited fashion, garnering interest and respect from his contemporaries and no doubt some stern glares from the prevailing establishment. While it may have been little, the commentary contains seven critical and visionary axioms.

Paraphrasing these in more modern terms, this is what Copernicus had to say about the cosmos:

- There is no single center to the universe.
- The Earth's center is not the center of the universe.

- The center of the universe is near the Sun.*
- The distance from the Earth to the Sun is imperceptible compared with the distance to the stars, and so no parallax is seen in the stars.
- The rotation of the Earth accounts for the apparent daily rotation across the sky of the Sun and of the stars, which are immovable.
- The annual variations of the Sun's movements across the sky are actually caused by the Earth revolving around the Sun.
- The looping (retrograde) motion that we see for the planets is actually caused by the movement of the Earth.

After this last idea Copernicus was sufficiently excited to add in his brief commentary: "The motion of the earth alone, therefore, suffices to explain so many irregularities in the heavens."

Here in these sentences was the genesis of a colossal revolution in human thought. Through the power of little more than deductive reasoning, Copernicus had set the cherished Earth spinning and traveling through the universe. But although the circulation of the *Commentariolus* helped him gain a considerable reputation, it wasn't until decades later that he finally took these writings and more thoroughly worked out the mathematical pieces of his theory in order to have them published—effectively posthumously—in the great *De revolutionibus orbium coelestium*, "On the Revolutions of the Celestial Spheres," in 1543.†

As much as this model shook the heavens into shape, it was also still very far from perfect. Despite, as we now know, correctly arranging the Earth, the Sun, planets, and stars in their respective

---

*You may wonder if this isn't inconsistent with the first axiom. Copernicus didn't mean for these to be a minimal set of statements; they're more like a laundry list of hypotheses.

†Exactly why it took him so long is a matter of some historical intrigue. One might speculate that a fear of having to battle the might of the church and establishment—itself reeling and testy with the impact of the Reformation—was a big factor in his uncharacteristic foot-dragging.

places, Copernicus still assumed certain properties that made fitting his model to astronomical observations awkward. In fact, rather than doing away with all of Ptolemy's complicated geometrical devices, Copernicus merely did away with some parts. He continued to use epicycles to get a better match to the real behavior of the planets and the Sun as they passed through their annual tracks.

The underlying physical arrangement was better, but the application of the model was still a bit of a nightmare, and the reason was that Copernicus was clinging to a set of ideas that went all the way back to Aristotle. His calculations assumed that all motions, whether on great shells or in epicycles, followed perfect circles and took place at constant velocity. This was consistent with classical ideas, wonderfully geometric, and—unknown to him—completely wrong. But Copernicus had certainly seeded the revolution of scientific thought, and what a revolution it would turn out to be.

❧

The decades that came after Copernicus's *De revolutionibus* brought a host of new dissenters from the Ptolemaic universe, as well as many equally vociferous defenders. Some of the dissenters, such as Giordano Bruno, paid dearly for their views. The Dominican friar was born in 1548, five years after Copernicus's death. His scientific and philosophical studies led him to advocate not only the heliocentric worldview, but also the idea that the universe was truly infinite, that the Sun was merely another star, and that there must be an endless number of other inhabited worlds across the immensity of existence. By building on the works of the ancient Greek atomists, Bruno advocated a prescient vision of nature. But together with his extremely provocative stance on other religious matters, it caught the full attention of the establishment, and in 1600 the Roman Inquisition burned poor Bruno at the stake for heresy.

During this same period the wealthy Danish nobleman and astronomer Tycho Brahe was making enormous strides in the observation and record keeping of the heavens. Without telescopes he used his keen eyes and clever measuring devices to track the cosmos—

inventing new versions of quadrants, sextants, and armillary spheres to measure angles, positions, and coordinate systems with impressive accuracy. One night in 1572 the twenty-six-year old Brahe witnessed a new star in the November skies of Western Europe. It showed no discernible parallax but had clearly not been there in previous nights, and Brahe came to the conclusion that the universe was therefore not immutable—it could change and it could change dramatically.

We realize now that he had observed a supernova, in this case the powerful implosion of a critically overweight white dwarf stellar remnant, some 8,000 light-years from the solar system. The experience of seeing this primal event helped encourage Western astronomers to devise even better ways of measuring the positions and brightness of objects and to explain their arrangements. Brahe himself worked hard to try to combine, or at least reconcile, the Ptolemaic cosmology with that of Copernicus. He produced his own "Tychonic" geo-heliocentric system, in which the Sun and Moon orbited the Earth, but all the other planets orbited the Sun.

As contrived as it was, he found this arrangement satisfying because he had still not detected a parallax for the stars, and keeping the "sluggish" Earth stationary meant he could easily account for that fact. Better yet, his system allowed a convenient compromise position for advocates of the Copernican vision who were still feeling very nervous about their scientific beliefs. But it was Brahe's meticulous care with astronomical observations that really set the stage for one of the most critical next steps—steps that came from his onetime assistant, the German-born Johannes Kepler.

Four years before meeting Brahe in 1600, Kepler had published a rousing defense of the Copernican system of the heavens in his *Mysterium Cosmographicum*—"The Cosmographic Mystery." Interestingly, Kepler was not only an obsessive mathematician, but also deeply religious, and he felt that everything that determined the positioning and motion of celestial bodies was of divine influence. (This might help explain why his first stab at modeling a heliocentric cosmos relied on a series of three-dimensional polyhedra nestling inside each other—a geometric and appealingly designed, but deeply flawed, concept.)

The full story of Kepler's life and studies is convoluted; he was an exhaustingly busy and productive person, particularly in matters of science. His investigations of optics led him to deduce the fundamental inverse-square law of brightness: the intensity of a light source is proportional to the inverse of the square of its distance. With the observation of another supernova in 1604, Kepler also reasoned, just as Brahe had, that in the absence of a measurable parallax, Aristotle's unchanging and immutable universe was probably not a correct model. But most important, when it came to the problems of the Ptolemaic and Copernican explanations of planetary motions, Kepler found himself in a unique position.

The high-living Tycho Brahe suffered an unfortunate and untimely death from an infection in late 1601, and Kepler wound up inheriting that master astronomer's most complete and precise tabulations of celestial positions and variations. Some sources suggest that Kepler was instrumental in making very sure that he got his hands on these records before Brahe's estate was disbursed. Having already started working with Brahe, he knew exactly what he needed.

Brahe's unprecedented measurements gave Kepler an opportunity to keep tackling the endlessly nagging issue of finding a perfect fit to the motions of planets, where solutions in the existing schemes were still full of holes, including so-called residuals, gaps between the predicted positions of planets and their actual places at various times. Planets would simply not be quite where models said they should be on particular nights, which was quite a glaring problem.

When Kepler settled down to study these extensive data, he chose to focus his attention on the observations of the planet Mars. That choice was, I think, one of the biggest strokes of educated luck in the entire history of Western science, even if it was likely encouraged by Brahe's earlier suggestions.

Of the six planets that Kepler knew of, Mars exhibited the worst residuals of all. In fact, Kepler demonstrated that Mars could not conceivably be following a fixed path if the Earth were at the center of everything. He went on to consider what had heretofore been absent from all models of the universe: the possibility that objects might not move with constant velocity all the time. By allowing this behavior into the mix he wrenched open a new window onto nature, because if objects moved with varying speed they might also move along paths that were not perfect circles. This wasn't an easy task—it took Kepler four years to get his answer, and another four years to publish it.

Kepler tried out various shapes for his planetary motions; egg-like ovals didn't work, nor did other forms. He then tackled the movements using only mathematics, obtaining a solution and rejecting it before coming back to the exact same idea by guesswork. The answer, he finally realized, was that all planetary motions belonged to a class of curves known as conic sections. These could be circles, parabolas, hyperbolas, and most critically—ellipses.

The reason why Mars has such awful residuals in the Copernican model is, as we now know, because it has the least circular, or the most elliptical, orbit compared with Venus, Earth, Jupiter, and Saturn. Of the planets familiar to Kepler, only Mercury had an orbit with greater ellipticity. But the observation of Mercury is complicated by its proximity to the Sun. Kepler deduced that in an elliptical orbit, the velocity of a planet or any object varies between its far and near point—from fast to slow. This variation was exactly what was needed to eliminate the residuals afflicting Mars.

Gathering his ideas together, in 1609 Kepler published *Astronomia Nova*, "A New Astronomy," to present the first two of his famous laws describing planetary motion: the path of every planet is an ellipse with one focus centered on the Sun; and a line joining a planet and the Sun sweeps out equal areas in equal time as the planet moves.

Kepler also realized that there might be some kind of unseen influence at play between the Sun and the planets (what we would today term a force). This concept was revolutionary, and although it was all couched in somewhat mystical terms, he went as far as suggesting that such an influence might be reduced with distance from the Sun. Hence the farther planets would move more slowly, which of course they do.

Merely a year after *Astronomia Nova*, in 1610, Galileo Galilei made his telescopic observations of the periodic motion of the brightest moons around Jupiter and of the phases of Venus. Both of these observations brought the clash between celestial worldviews to a boiling point by providing even more convincing evidence for a Sun-centered system, putting Galileo into a head-on collision with the established doctrines of the time. But there is something else lurking in the picture of the universe that emerged out of Kepler's work that is just as vital to our quest to understand our cosmic significance.

If planets follow elliptical paths as a general rule, and those paths need not be all within a single plane around a centrally massive star, the possibility exists for an extraordinary range of planetary motions and arrangements that nonetheless all obey Kepler's rules (and what would soon be Newton's physics). I doubt anyone suspected it at the time, but the door had been opened to a universe of far greater abundance and diversity than anything yet imagined, even by the atomists and pluralists of the past. Galileo's observations produced other surprises as well. With a telescope he could find stars that were too faint to be noticed with normal human vision. When he looked at the seemingly smooth and cloudy expanse of the Milky Way, he was amazed to discover that it was in fact *made* of stars, so many and so tiny that the naked eye blurs them together. His observations of these other phenomena tend to get less attention than they deserve, but they were beginning to reveal the true enormity of nature.

Just like the shock of Tycho Brahe's supernova, the notion that there were hidden objects in the sky ran counter to the cosmological perceptions of the time. These observations, together with Antony van Leeuwenhoek's discovery a few decades later of the teeming microscopic universe in every speck of water and in human sputum, began to lift a previously opaque veil from the enormous intricacy and depth of reality. Yet these pivotal revelations of the sheer depth of nature—inward and outward—didn't cause anything like the controversy sparked by the simple decentralization of our place in the universe.

The upset of that shift was mostly confined to the camps of the church and establishment. In fact, it doesn't appear that either

Galileo or Kepler saw heliocentrism as a *demotion* of the terrestrial status. Quite the contrary: it meant we were no longer at the "bottom" of the planetary pile; we were on a noble orb within the pathways of others. Ironically, Kepler even wrote that he considered this to mean that the Earth was at the *center* of the planetary globes (the orbits), with Mercury, Venus, and the Sun on the inside, and Mars, Jupiter, and Saturn outside. Yet again, such a robust certitude about our significance in the grand scheme of things actually lessened the impact of the growing evidence for the true immensity of nature, from the world within to the world without.

Time passed, and the year 1642 saw the death of Galileo in January, and the birth of Isaac Newton in December. The full story of Newton's life, much like those of Copernicus, Bruno, Brahe, Kepler, and Galileo, is immensely rich. But the most important piece for our quest comes with the publication in 1687 of his monumental *Philosophiæ Naturalis Principia Mathematica*—"Mathematical Principles of Natural Philosophy," more often known simply as the *Principia*. In this text Newton lays out not only the mathematical laws of motion, including the concepts of inertia, momentum, force, and acceleration, but also a universal law of gravitation.

Newton saw that the attraction of bodies to each other could be described as a force that grew with mass but diminished as the inverse of the square of distance. From this hypothesis he derived the mathematical proof of Kepler's empirical laws—showing for the first time that the rules governing the planets came from fundamental physics. He also presented analyses of the motion of the Moon, the paths of comets, and the gravitational interactions of more than two bodies. He noted that despite the clear heliocentric nature of the solar system, the Sun does in fact orbit around a variable point—the center-of-mass or balance point of all objects in the system. He even determined that this point was close to the observed surface of the Sun—well offset from its core, a result largely due to the gravitational bulk of Jupiter and Saturn. (This latter fact is familiar to modern astronomers, because the same type of offset in other star systems provides one of the key techniques for finding exoplanets, the planets beyond our solar system that we will encounter in a later chapter. We measure the orbital motion of a star around this pivot point, since it marks the presence of unseen but massive worlds.)

Newton was a strange and complicated character with deeply held religious beliefs, and for him this beautiful physical explanation of planetary motion was evidence of a supreme divinity, maintaining the paths of objects in a perfect clockwork dance. For other thinkers in the following century, like the great French mathematician and scientist Pierre-Simon Laplace, it meant quite the opposite. There need be no guiding hand, no preordained paths or configurations in the Copernican universe, just the innate physical laws to determine where and when any object would find itself. But Laplace also felt that, armed with these laws, and with a complete knowledge of the locations and movements of all objects at any time, we could always know the past and future. There might be no guiding hand, but there was determinism in the universe.

Observations of the cosmos around us continued to improve as the next few hundred years went by, as did the toolbox of mathematics and physics available to work with. Mystical and philosophical reasons for the arrangements of nature gave way to the application of simpler, more general laws. At the same time, what we knew of the composition of the universe became richer and richer, and the notions of extreme scale and of the variety of phenomena that were hidden by time or by faintness grew and grew. The idea that stars were not only vastly distant, but perhaps scattered throughout an enormous volume, gained greater acceptance among philosophers and scientists. With this growing sense of scale, even the musings of the ancient Greek atomists on an infinite cosmos returned to people's thoughts.

Our scientific sense of our own importance also evolved, in a variety of directions. Hot on the heels of Newton, the Dutch scientist Christiaan Huygens wrote about his thoughts on the possibility of extraterrestrial life just prior to his death in 1695. Huygens was convinced of the "plurality of worlds," imagining a wealth of watery and hospitable locations in his telescopic observations of the planets, and even of the moons of Jupiter and Saturn. Life like ours seemed to him to be nearly inevitable elsewhere. This was certainly not a view shared by everyone, and debate on our place among the stars raged on.

Something else was also taking place during this period: a contentious and surprisingly underappreciated scientific debate that began in the early 1700s and arguably didn't reach any kind of satisfactory closure until as recently as the 1970s. With the monumental advances in physics due to scientists like Kepler, Galileo, Newton, and Laplace, the solar system became a phenomenon whose origins now begged for a proper scientific explanation.

Where did the Sun and planets come from if they were less of a divine construction and more of a consequence of the laws of nature? The answer, as I'll soon show you, is quite stunning, and neatly frames our more modern debate on origins and significance. Before that, though, we need to reach the present day in our brief history of cosmic perspective.

By the end of the 1800s we were beginning to appreciate the true vastness of the universe. Stars were now accepted to be extremely distant analogues of our Sun, a fact supported by astronomers finally having success in measuring their barely noticeable parallax movements from Earth's annual motion through space. New planets had also been discovered in our solar system—from the darkly distant Uranus and Neptune to minor but still massive objects like Ceres and Vesta, just beyond the orbit of Mars. And the elemental composition of extraterrestrial objects was beginning to reveal itself through the spectra of light, including the discovery of an atomic species in the Sun—the stuff we now call helium.

But other big questions remained: Was the universe infinite in space or perhaps even time? Was the spread of stars we call the Milky Way the full extent of the universe, or could some of these other strange little smudges of nebulosity, like the one called Andromeda, actually be other "island universes," other galaxies?

In an unprecedented burst of discovery and invention, in the first three decades of the twentieth century there was another series of scientific revolutions. Their stories have been told innumerable times: Albert Einstein's theory of relativity, the measurement of the true scale of the cosmos and the nature of galaxies, and the development of quantum mechanics. These all produced radical views of nature that dealt with the intertwined properties of the very large, the submicroscopic, the fast and the energetic, and the underpinnings of reality itself. These revolutions would also have to confront, and contend with, our perceived place in the cosmos.

The implication of a heliocentric Copernican model was that the universe would look more or less the same no matter which planet you stood on. The obvious extension was that the universe would look more or less the same *wherever* you stood within it— from our solar system to another, or from our galaxy to another that might be tens of millions of light-years away. For Einstein, working in the years following 1915, this was a philosophically comfortable proposition, and made the application of his theory of general relativity to the universe as a whole that much more straightforward, giving birth to the so-called *Cosmological Principle*.

In slightly more technical terms, the idea stated that the universe was homogeneous. While it might contain many small asymmetries, like patches of stars and galaxies, it would have the same amount of these lumps and bumps no matter where you were. It's a bit like the Earth's terrain: some places are mountainous and some places consist of flat oceans, but on average you can always find roughly the same mix of mountains and oceans wherever you are. This was very helpful if you were trying to apply a generalized theory of space and time, as Einstein was, to the workings of the cosmos.

It also meant assuming that the universe was isotropic, meaning that it would look the same in all directions from any place. This is a bit harder to swallow. After all, we can hardly claim that this is the way we experience the world or the solar system, and even the interstellar night sky is beset by big nonuniformities like the band of the Milky Way. But again, on scales that reach beyond our galaxy and out into the cosmos, the number and arrangement of objects seen in any direction should be more or less constant.

The first time that anyone seems to have publicly connected this cosmological principle to Copernican ideas was in the early 1950s, when the famous Austrian-born physicist Hermann Bondi used the phrase "Copernican Cosmological Principle" in his discussion of a now-disproven cosmological model known as the steady-state theory.

Much as its name implies, the steady-state theory proposed that the universe was eternal, with no beginning or end. To help make this theory palatable, Bondi asserted an even stronger principle: that not only would the universe appear the same in all directions to any observer anywhere, but for any observer *at any time*. Although we now know that our universe is most definitely not in a steady state, the Copernican Cosmological Principle reinforced the general notion that there was absolutely nothing special or privileged about our place in the cosmos, throughout space or time.

This middle period of the twentieth century saw the explosive development of a multitude of fields, from cosmology to microbiology and genetics, as well as the emergence of several generations of extremely influential scientists. But as it became clearer and clearer that the universe itself was an evolving and diverse place, several different people had begun to notice certain strange coincidences in the value of fundamental physical constants. These are numbers that describe things like the strength of gravity or the masses of subatomic particles, and in particular the estimated lifetime of the cosmos. Certain combinations of these numbers could yield surprising relationships. For example, the ratio of gravitational and electric forces, involving constant quantities describing the strength of gravity, and the masses and charges of electrons and protons is about 10^{39}. This number is remarkably similar to the current age of the universe when described in atomic units of time (a unit being about 2×10^{-17} seconds), a fact first pointed out by the physicist Paul Dirac.

But why should these immutable constants be related to the age of the universe *now*? Too far back or forward in cosmic time, this obviously wouldn't be the case. Furthermore, at some other cosmic time, conditions might not allow for any intelligent life to be around to observe these coincidences in the first place! This was a rather pesky issue for a Copernican Principle, since it suggested there was something special about when and where we found ourselves, and about the present physical properties of the cosmos.

The final, definitive proof that the universe was finite in age came in 1965, with the discovery of an all-pervasive flood of microwave photons originating from the young cosmos—effectively part of the remains of a hot big bang. This trace of a very different universe, one that was once upon a time ferociously dense and energetic, was more than one fly in the ointment. It was a whole bucket of flies, dumped into the vat of Copernican mediocrity. And things came to a head in a famous presentation by an Australian-born physicist named Brandon Carter in 1973.

Carter, who has played a central role in the modern development of black hole physics, was encouraged by the interest of a number of colleagues, including the physicist John Wheeler and a young Stephen Hawking. So he chose to stir things up at no less than a special conference in Krakow, Poland, held to commemorate the five hundredth anniversary of Copernicus's birth. In his talk, Carter articulated the ideas that had been brewing among a number of scientists puzzling over all these apparent coincidences between cosmic properties and our circumstances. He plunged into the thick of it by discussing just how different the universe might be if just a few characteristics were changed—like the relative strength of the fundamental forces that glue matter together.

Considering these changes raised an intriguing idea that Carter elaborated on for his audience. A tweaked version of nature might, for example, make no stars, but since we come from the elements produced by stars, and since we are here observing the cosmos, this very fact can be used to tell us something about the universe we live in. In other words, our existence itself tells us something about the nature of physics in the universe—we might be more important than we thought. Carter labeled this approach to examining the cosmos as the "anthropic principle," since "anthropic" means that something pertains to human existence. This wasn't exactly what he was driving at, since it could be *any* observer of the universe, not just humans. But although he later proposed a more semantically accurate term, the word "anthropic" stuck.

The underlying sense of this approach to understanding the world is well summarized in Carter's own words at the time: "Copernicus taught us the very sound lesson that we must not assume gratuitously that we occupy a privileged *central* position in the Universe. Unfortunately there has been a strong (not always subconscious) tendency to extend this to a most questionable dogma to the effect that our situation cannot be privileged in any sense." The point being that one cannot, and should not, ignore the multitude of phenomena that apparently need to line up just right for life, and us, to exist.

Now, a great deal has been written about the anthropic principle. It's been a veritable gold mine for some physicists and many philosophers, and an often confusing and confounding topic for snake-eats-own-tail conversations over cocktails. Extreme versions of the principle have even been constructed to argue that a viable universe *must* produce intelligent life capable of observing it—a notion that I'm going to steer very clear of.

However, the anthropic principle is an important idea, one that prompts us to face some of our preconceptions about the cosmos around us and to examine our innate biases of observation. And since it directly challenges the Copernican Principle (or rather, what has come to be the orthodoxy of our mediocrity), we ought to look at a couple of the details.

These days, anthropic ideas tend to crop up mostly in discussions of a phenomenon called "fine-tuning," which involves a more detailed examination of the cosmic coincidences that originally led to scientists' puzzlement over these questions. The notion of fine-tuning goes as follows: if we look carefully at a variety of properties of the universe, embodied in constants of nature such as the strength of gravity relative to other forces, or the proportions of matter and energy in the universe, we can see that were these properties to be changed by a small amount, life might not have arisen.

Except this tweaking is a little more complicated, because what we really mean is that objects like stars and galaxies wouldn't exist, or that they'd never forge the heavy elements, like carbon, that are so critical for the chemistry of life. So, in other words, a variety of primary cosmic functions would fail to set the stage for the secondary ones that we rely on. This, of course, also presumes that life has to be like us—but it does seem hard to imagine how a universe of just hydrogen and helium could give rise to structures with the complexity seen in carbon-based life.

Exactly which characteristics are most important to life's existence is not immediately obvious. The best way of narrowing down the possibilities comes from making clever mathematical combinations of quantities that in turn relate to tangible phenomena. The scientists Bernard Carr and Martin Rees did just this in 1979, and later on, in 1999, Rees revisited the ideas and came up with six numbers that have to lie within a comparatively narrow range for our universe to be the way it is, and to be amenable to life as we know it. These numbers are:

- the ratio of the strength of gravity to electromagnetic forces
- the percentage of matter that's converted to energy by the nuclear fusion of hydrogen into helium
- the total density of normal matter in the universe
- the energy density of quantum vacuum fluctuations (which may be the same dark energy that is accelerating the expansion of our universe)
- the sizes of the minute variations in the early universe that eventually grew into such structures as galaxies and their groupings
- the actual number of spatial dimensions in our universe

It's quite an array, and the odds appear extraordinarily low for any universe popping into existence to—by chance—have all the necessary properties for life to arise. Of course, as you read this you may be thinking to yourself, "But if it weren't like this we wouldn't be around to think about it; we simply have to exist in this type of universe." And that's absolutely correct. However, if this is the one and only universe, with no universe before or after it, that raises the uncomfortable question of why it turned out this way: suitable for life.

One of the most appealing answers is that ours is only one of a near-uncountable array of viable universes. It is a single example of a type of reality that is separated by time and space, or dimension, from gazillions of others. The word "appealing" may seem almost comical here—I've just invoked what you might think is an unsub-stantiated hypothesis for the nature of reality. But this idea of a "multiverse" is a leading contender for the deeper truth of the phys-ical world. Indeed, when Brandon Carter came up with the anthropic principle he was already thinking along these lines.

Although I don't think anyone could yet claim that we have direct evidence for there being a multiverse, there are several compelling theoretical ideas that readily accommodate it, and that also seem to provide solutions to other aspects of fundamental subatomic physics and cosmology. If correct, it would mean that there is no fine-tuning problem per se. We simply exist in one of the universes that happen to be "right" for the formation of galaxies, stars, heavy elements, and complex carbon chemistry. It sounds as if this would neatly resolve the issue, and in many respects it would—if we actually knew that we lived inside a multiverse.

Another tricky thing about the multiverse solution is that it is still partially motivated by the notion that our particular universe is genuinely fine-tuned for life. It's still thinking in pure anthropic terms—and in those terms it is assumed that life is entirely repre-sented by us. One doesn't need to invoke any other life, or type of life, anywhere else in the cosmos to make this argument, and that seems a bit parochial. It would be like arbitrarily basing your entire philosophy of science on the existence of a particular type of un-usual parrot. The last thing we want is to be misled up a dead-end alley. So it's worth pursuing this further, because we don't yet know if we live in a part of a multiverse, and because none of the above gets us much closer to evaluating our immediate cosmic significance, or lack thereof.

With only some very simple changes to our perspective on the universe, one can appreciate how some aspects of fine-tuning and anthropic reasoning begin to look like a bit of a distraction in our quest for significance. I'll visit some other ideas along these lines, but let's kick off with the following—a playful question to make a serious point.

Let's suppose for a moment that Galileo Galilei's interpretations of his observations of the universe had been immediately embraced as a crowning achievement of reason and technology. Instead of being raked over the proverbial coals, he becomes the darling of the seventeenth-century church and state. And in this alternative timeline the enlightened establishment seizes this moment to initiate a grand technological push, seeing the potential economic benefit of engineering and science.

Flush with the warmth of acceptance and patronage, Galileo quickly sets to work building sophisticated telescopes that will let him become the first human to find planets around other stars, and to confirm the presence of biological systems across many of these worlds. It's a lovely fantasy, a horse- and water-powered science-fiction reworking of history, but most critically it also lets us ask how things would be different *today* if this had really happened.

For the intervening centuries we would have known that life was not confined to Earth alone, and we might even have figured out if any of it was more than just microbes or uncommunicative creatures. In either case, the point is that we would have at our fingertips a real answer to the question of how likely, or how unusual, our type of life is in this universe.

Let's suppose that in this parallel reality we find that life resembling Earth's is moderately common. It occurs often, but it's neither spilling across every suitable world, nor so unusual that it exists only in certain galaxies sparsely distributed across the universe. What then of the fine-tuning arguments embedded in the anthropic ideas about cosmology? It might not even occur to us to ask these questions in the first place. It would be like suddenly deciding to question why the world produces a certain number of snails. But even if we did ask, the notion of "tuning" doesn't exactly hold up so well in this hypothetical reality.

The universe would appear to be suitable for producing only *some* life, hardly the stuff of great cosmic significance; a modestly fertile pond of occasionally making something functional. Now of course we could also discover that the answer lies at one of two possible extremes: from life as an utterly freakish rarity across 14 billion years' worth of cosmic time, to life run amok, clogging every planetary system with some new variant.

In the former case we'd hardly think the universe was well suited to life, and the coincidence of physical parameters with the requirements of life would just be seen as a cruel joke. By contrast, in the latter case we might deduce that life itself, not so much the cosmic underpinnings, is a remarkably robust phenomenon. We might even be asking whether there were any (almost unimaginable!) circumstances in which life could *not* spring from the underpinnings of the physical laws.

There are two points to make here. The first is trivial, and it is that the questions we end up asking are themselves a direct function of what we've already observed about our surroundings. The second is much more important, because unlike the inhabitants of my fantasy Earth with its alternate astronomical history, we do not know at present which of the above scenarios applies in this universe.

Furthermore, fine-tuning may not be a make-or-break situation. Instead it could be a "coarse-tuning" problem with the real fine-tuning hidden inside. As in my fictional example, the issue of the universe being suitable for life is not an all-or-nothing question. It could rest within a spectrum of fertility and likelihood. In fact, I think that there is an implicit assumption in anthropic arguments that life is a bit wimpy, that it must have everything perfectly aligned or else it won't happen.

Yet we know from the abundant and remarkable wealth of paleontological evidence on Earth that brutal natural selection has allowed life to fine-tune *itself* to the environment around it. In the face of varying chemical mixes and abundances of vital elements, as well as a multitude of different sources of energy, life has found a way. Admittedly this is within a set of circumstances determined by the basic laws of the universe. But life on Earth has become diverse enough to exploit a variety of secondary biochemical strategies—not just a single one.

It's not obvious that life needs anything more than a rough-and-ready environment to originate and survive in. So true cosmological fine-tuning should be more about the particular ease with which life can occur—and for now, at least, I make no distinction between intelligent life and "simple" life, since there's nothing simple about life in any form.

This way of looking at things is consistent with studies of the coincidences of physical constants and other quantities such as the proportions of mass and energy in the universe. In most of these cases there is a little bit of wiggle room—an issue well illustrated by the way in which elements are produced by nuclear fusion inside large stars.

During the first half of the twentieth century, scientists realized that conditions inside stars could give rise to the fusion of atomic nuclei, powering their prodigious energy output and forging heavier and heavier elements. But the recipes were not straightforward, and in the early 1950s the English physicist Fred Hoyle realized that there was a problem with carbon. At the time, physicists' emerging theories of stellar fusion were suggesting that stars should make relatively little carbon. But Hoyle observed that since we're made with carbon, the universe must actually have a way to generate plenty of it. This puzzling discrepancy helped prompt him to find that carbon-producing process.

He discovered that carbon gets readily formed in the universe because of a specific phenomenon. The energy of one of the stages involved in combining three helium nuclei in a star's interior almost exactly matches that of an agitated carbon nucleus—which is the natural product of combining those three helium nuclei. This correspondence results in what's called a nuclear resonance, a harmonizing of energy states that boosts the efficiency of the nuclear reaction enormously, so that, instead of stars making almost no carbon, they can make lots of it.

For a long time, the carbon resonance was considered one of the strongest pieces of evidence for an anthropic principle to be in play—namely, the existence of carbon and carbon-based life itself suggested this special nuclear process in stars. This is true, but only up to a point, because there is a devil in the details. We now know that these nuclear energies need not match so precisely for carbon to be produced: there's a certain amount of leeway, and so the fine-tuning is not quite so fine after all. And the same is true of many of the fine-tuning parameters. Things could be a teeny bit different and conditions would still be passably okay for life as we know it.

The concept of such wiggle room goes deeper still. If we are able to eventually *measure* the propensity of the universe for making life—the efficiency, or density, with which life occurs in any given patch of the cosmos—we will have a new tool for probing the basic properties of nature and for *predicting* the occurrence of life according to those fundamental circumstances.

This is not to say that there is necessarily something "special" about life but rather that life is an excellent example of a highly complex phenomenon, and one that is conceivably the most complicated in the cosmos, with an intricate web of connections to many key features of the physical laws in this universe. As such, life represents a natural litmus test of cosmic properties, a canary in a cage for examining the detailed interplay between characteristics where there are a vast array of potential permutations and combinations.

This is more than rephrasing the anthropic argument. At its core that argument states that a sole occurrence of life yields predictions about the universe. Instead I'm proposing a way to learn how to take the properties of the universe and predict the abundance of life, and therefore predict our significance. It's a bit like taking an opinion poll and using that to predict the outcome of an election.

The catch is that we have developed a bit of a complex thanks to Copernicus, whose ideas so clearly and accurately describe our solar system, and who helped break us out of a deep and awful rut of provinciality. The apparent confirmation of our unprivileged ordinariness is surprisingly compelling (flying as it does in the face of all our solipsistic and egotistic tendencies), and it has allowed us to make extraordinary progress in understanding the universe around us, as well as the universe within us. But it creates some confusing situations.

On the face of it, the Copernican Principle suggests that we *cannot* be alone in the universe; we are neither central nor special, and our circumstances should be representative of the circumstances in any number of physical locations at this point in the history of the universe.

So, by that logic, not only should there be plenty of other life out there, but a great deal of it should be very similar to that on Earth. But is the assumption of our own averageness really a sound basis for making such an argument? It smacks of an overly literal reading of the scientific gospel. Copernicus was simply trying to understand the motions of planets in our solar system in the least contrived and most mathematically logical way. Are we reading too much into what was primarily a mechanical solution to a mechanical problem?

Recognizing the limitations of the Copernican Principle is not a particularly controversial suggestion. Anthropic ideas are one good example of a counterpoint, and many astronomers and physicists find similar clues in some straightforward aspects of our circumstances. The fact that we are so manifestly located in a specific place in the universe—around a star, in an outer region of a galaxy, not isolated in the intergalactic void, and at just this time in cosmic history—is simply inconsistent with "perfect" mediocrity.

The situation is this: a Copernican worldview at best suggests that the universe should be teeming with life like that on Earth, and at worst doesn't really tell us one way or the other. The alternative—anthropic arguments—require only a single instance of life in the universe, which would be us. At best, some fine-tuning studies suggest that the universe could be marginally suitable for heavy-element-based life-forms, rather than being especially fertile. *Neither* view reveals much about the actual abundance of life to be expected in our universe, or much about our own more parochial significance or insignificance.

And we want some answers! So to find the truth, we need to take a good, careful look at the nature of the multifaceted array of matter in the universe around us and within us. We need to navigate a path somewhere between the assumptions of mediocrity and the assumptions of fine-tuning and an anthropic worldview. We need to figure out a way to see around these extremes and to make actual measurements of what we find.

The story in *The Copernicus Complex* is about the great adventure and unfolding meaning of that effort to discover the universe within and without. It is also about our past and future—especially the future. But more than anything else, it is about that deep-rooted need, that frustrating yet recurrent itch that each and every one of us gets when we try to think about our place in the grand scheme of creation.

We need to *know*, to truly know, whether or not we are significant—not just emotionally or philosophically, but objectively, in cold, hard facts and figures. It's one of the greatest scientific challenges facing us. Part of that challenge is to understand and see past our intricate models of the world, which serve us so well but time and time again need revision, updating, and sometimes discarding. So the next step to take is from the familiar Earth of today to the unfamiliar Earth of yesterday and tomorrow. If we want to place ourselves in context, we need to begin to reach up and out to cosmic time and space, as well as down into the microcosm. We'll find that what the enterprising scientist Antony van Leeuwenhoek saw unfolding in his microscopes more than three hundred years ago was merely the beginning of a truly fantastic voyage.

2

THE TEN-BILLION-YEAR SPREE

Earth's changing geographical zones can be awe-inspiring. If you want a good example, take a trip by road up into what the Chileans call "the hills," at the southern end of the vast Atacama Desert in South America. If, like me, you want the maximum effect, start your day at the Pacific Ocean, where it comes rolling and crashing onto the beaches of La Serena, some three hundred miles north of Santiago.

Here I'm rudely awakened by the raucous cawing of hungry birds as they swoop through the moist and salty air, thick with the odors of seaweed and deep oceanic life. Out on the beach a few lone souls jog along the sand as the morning sun begins to burn away a light mist. It's the start of a daily cycle of evaporation and condensation that's been playing out for millions of years along this shoreline. For my nose and lungs it's an aperitif bubbling up directly from the planetary biosphere, and I draw in deep cool breaths before starting the drive along the dusty roads that head inland toward my final destination.

Farms and vineyards full of lush green tones come and go as I make my way through the great Elqui Valley, a huge V-shaped incision reaching far inside the continent. Grapes and tropical fruit are among the principal crops here. I can see why; the sunlight is brilliantly saturated, and the huge valley oozes with life. It's a rich and fertile incubator basking in warmth and energy.

Big water towers dot the edges of verdant fields, emblazoned with the larger-than-life labels of the national brands of *pisco*, a potent grape brandy that's been made in this area for almost five hundred years and consumed up and down the country.

But things alter quickly as I move farther inland, undeniably climbing upward from sea level. A huge dam, completed only a dozen years ago to improve crop irrigation, looms in the distance. Its colossal embankment of rock and concrete bridges the valley from side to side on a scale so grand that's it's hard to take in, reworking the natural sculptures of geophysical time.

Soon I'm turning away from these scenes filled with human purpose, and the green growth around me changes rapidly to a mix of scrubby vegetation and brownish-red rock and dirt. A few minutes more and it is as if I've traversed an invisible border, passing into an entirely different place, a mineralogical zone removed in time and space.

These are the real hills, except they're not hills by any paltry standard that I'm used to. They're part of the incredible concertina of Earth's crust that makes up the Andean mountain range, a truly monumental geophysical feature more than 4,000 miles long, pushed skyward as oceanic basalt dives underneath the floating South American Plate of continental rock. Here the tortured constrictions of a cooling planet are writ large. An ever-tightening outer skin of crystallized material sits raftlike on an ocean of magma, where it cracks into colossal lithospheric plates of solid crust that jostle for equilibrium within a deep well of gravity.

The road begins a gentle spiral as it climbs higher and higher, and the land becomes increasingly dry and empty. Progress is slower, as the occasional rockfall has scattered sharply angled pebbles and dirt onto the path. But eventually I spot my destination in a flash of reflected sunlight atop a great rise. Here are the pure white and silver coatings of tall telescope domes sitting amid an endless blue sky. These modern versions of spires and steeples belong to the Cerro Tololo Inter-American Observatory, my home for the next week.

I'm here to perform a comparatively mundane astronomical duty: to take calibrating snapshots of a few dozen distant cosmic islands—a set of otherwise nondescript galaxies strewn hither and thither across the visible universe. To do this, I have one of the telescopes all to myself for several nights; I will huddle next to this instrument in a snug room packed with computers and monitors.

From inside this little cave I can control the dome's machinery and the telescope's sensitive digital camera, whose innermost pieces are chilled with regular infusions of liquid nitrogen—a task that tests the coordination and nerves of even the steadiest hand in the depths of southern darkness. The images I hope to capture represent just one stage in a lengthy project of mapping and measuring these remote stellar townships, tracking their gentle evolution across cosmic time, an effort that will occupy my colleagues and me for many years.

Like any professional observatory, Cerro Tololo runs on routine. During the day its technicians and engineers fix, clean, and test the telescopes and their attached instruments. During the late afternoon the nocturnal astronomers emerge groggily from their hillside dormitories—looking for food before spending a night on the tiles. And every evening after supper they wend their way up to the top of the mountain.

Here the peak was leveled with dynamite and heavy machinery in the 1960s to accommodate half a dozen large telescope domes and equipment. It's a serene and rather beautiful monument to human curiosity and accomplishment, perching on what really does feel like a gateway to the skies. This evening is no different, and soon I am in my place, switching on devices and fumbling with liquid gases before opening the curved body of my telescope's dome to vent the hot air from a day's worth of beating light.

Every astronomer has his or her own habits and unwritten traditions about using telescopes. For me an important one is watching the sunset. It's not for any particularly romantic reason. I like to get some air before settling down to a long night of work, and I like to get a firsthand sense for what the sky is *feeling* like and what the local weather seems to be doing—two important clues to the quality of the data I hope to collect.

At the top of Cerro Tololo it's easy to do this; you just walk outside and crunch your way across the gravel on the shorn-off mountain-top. At the edge of this ridge the ground drops precipitously away, leaving a magnificently open view of the distant landscape and soaring heavens.

A few other astronomers have taken up similar perches along the ridge, each standing like a philosophical meerkat inspecting its kingdom. Off ahead of us on the western horizon, the undulating hills form a wavy silhouette between sky and land that casts shadows across the desert as the Sun lowers out of sight and darkness spreads across the world.

Except this evening, as the Sun vanishes and the cloud-free canopy above us begins to dim, it doesn't look like any sky I can remember seeing before. From the horizon's edge where the Sun has set, up toward the zenith above, is an angled glowing band

Figure 3: The zodiacal light seen a few minutes after sunset from another Chilean astronomical mountaintop, that of La Silla Observatory on the outskirts of the Atacama Desert at an altitude of 8,000 feet. (European Southern Observatory, 2009, Y. Beletsky)

that tapers away as it climbs. It's like the great blade of some un-earthly luminous sword. It is far too bright to be the stars of the Milky Way.

I'm startled and rather puzzled, and I sidle over to one of the other astronomers standing quietly like me to witness the end of the day. I point, I mumble my confusion, asking for an explanation. And with two simple words he answers me.

This glow in the heavens is something I should've recognized, but it is usually washed out in any but the darkest of skies, far from civilization. It has also become lost in the recesses of my memory along with the yellowed pages where I'd first encountered it; the zodiacal light, part of the ethereal corpus of our solar system itself, and a signpost to the origins of everything around me.

❀

Where we live is a pretty open place, measured by local celestial standards and in human terms. From where you're presently sitting or standing, it's about 240,000 miles of vacuum to the Moon. From there it's about 93 million miles of interplanetary void to the Sun, a distance that light slogs across in about eight minutes.

The great Sol, our churning sphere of fearsome nuclear energy, is itself an imperious 865,000 miles across. But between the Sun and the outermost major planet, Neptune, is a staggering gulf of about 2.8 *billion* miles, on average. By comparison, the sizes of the planets range from 89,000 miles in diameter for the gas giant Jupiter, all the way down to about 3,000 miles for the dense rock of Mercury. Thus, what to us are entire worlds are barely more than specks to the cosmos—mere crumbs whizzing around a modest stellar candle in a cavern of space.

These dense little planetary bodies orbit the Sun in paths that lie in a similar plane—in fact, taken together their movements sketch the rough outline of a single great disk. Lots of other far tinier bodies also occupy this same region and beyond, out to the farthest reaches of our solar system. Trillions of solid pieces of rock and other, frozen compounds orbit the Sun, ranging from millions

Figure 4: The Sun and major planets of our solar system shown to approximate scale. The Earth is the third speck from the lower left. (NASA/JPL/Space Science Institute)

of rocky asteroids that are kilometers across, to an unknown but enormous number of smaller boulders and pebbles of solid material.

Not all of these small bodies stick close to the disk of planets; many are on orbits that tilt away from this platter. There are also the ice-rich masses of cometary nuclei. These might look like asteroids if given a cursory glance, but these nuclei can flare into great glowing tails of luminosity if they stray too close to the Sun and begin to boil off their volatile mixtures of water and other chemicals.

Many millions of the smaller bodies are located in the asteroid "belt" between the orbits of Mars and Jupiter, yet this zone is so huge that these objects are sparsely distributed by our Earthly standards. There are gaps of millions of miles between the larger ones, and our spacecraft can fly through the belt with nary a chance of collision. Other families of lumpy solids are more like errant bugs, flying in and out of the solar system's disk in all manner of configurations.

Yet farther out from the Sun are more asteroids, minor planets, and even dwarf planets (of which Pluto is one, surrounded by its five moons). These occupy orbits in the regions between Jupiter, Saturn, Uranus, Neptune, and beyond—where they join a family of so-called trans-Neptunian bodies. These exist within a frigid and still-mysterious zone known as the Kuiper Belt, which itself reaches about fifty times farther from the Sun than Earth's orbit.

Figure 5: A true-to-scale schematic of our solar system. *Top*, the correctly oriented orbits of the inner planets around the Sun, zooming out (*center*) to the orbits of Jupiter, Saturn, Uranus, and Neptune, and Pluto's larger tilted trajectory. All of this is ringed by the outlying, torus-shaped Kuiper Belt, which in turn lies within the vast, icy shell of the Oort Cloud (*lower right*). From the edge of this cloud the nearest star is a distance of about 3 light-years, some 18 trillion miles.

Out here sunlight is two and a half thousand times fainter than we experience it on Earth, and our nurturing star is reduced to a bright point in an otherwise eternal night. Beyond this zone is a still speculative region we call the Oort Cloud, hundreds to thousands of times farther again from the Sun. We think it's the source of certain types of comets, those on such long orbits that they may only appear every few hundred, thousand, or even million years. To produce the comets we see, there must be trillions of icy objects in this enveloping Oort borderland, occasionally nudged inward to fall toward the familiar planets.

This is the loose detritus from the deepest history of our solar system, and possibly even the distant staging post for icy interlopers from other passing stars, encountered during our continual orbital circuit around the Milky Way galaxy. Beyond this point, maybe as far as a light-year from the Sun, is true interstellar space and the beginning of the rest of the universe.

It's a vast, skeletal array of points, mostly emptiness. But there's something else as well that fills the interstices of our solar system ever so slightly, a component of interplanetary dust. Tiny grains of silicates and carbon-rich material are spread in an enormous and tenuous haze that blankets the inner planets. Distributed in the form of a puffed-up disk, this cloud reaches from around the orbit of Jupiter to within that of Mercury.

At their largest these grains are only a tenth of a millimeter across, barely bigger than microscopic, and they number no more than one in every cubic kilometer. But the solar system is a very big place, and a colossal number of these particles spread across local space can scatter and reflect light just as if they were motes glimmering in a sunbeam as they float across a room.

Standing on the mountaintop in Chile, I'm seeing the glimmer stretched across the sky. Photons of light from the Sun have careered outward into the solar system, only to be scattered off dust grains and sent on a new path that leads directly into my retina.

Ancient Islamic astronomers called this glow in the heavens the "false dawn," since it can also appear to the east an hour or so before the Sun rises—as if time itself disappears and the Sun returns early to light up the world again. In fact, it's not so much the world that's being illuminated, but the framework of the solar system, a dusty impression of the alignment of all the planets in their huge disk of orbital paths, and of all the multitude of other objects sharing this same space. It is spectacular.

All of this glowing dust also shares a common ancestry with the types of solid material that once produced the Earth itself. That stuff was coagulated and accumulated, melted and refrozen, eventually morphing into the layered minerals of the cores and rocky crusts of planets and moons. There is a clear lineage shared with the substances that I had driven across and through on my way from the Pacific Ocean to the Andes. These very same compounds and elements help fertilize the soils of the Elqui Valley, and the same components shift as gravel beneath my feet. It is also, I realize as I stare up at the glow of quadrillions of distant dusty grains, precisely the same kind of stuff that I'm made out of.

It's a profound moment for me, an unexpected reminder of the deepest connections between my brief existence and Big Important Things. But how did the chain of events leading to this instant really come about? How do grains of space dust like this turn into planets? How do they form worlds that harbor oceans, mountains, and living, breathing creatures asking about their cosmic significance?

The history of the Sun, Earth, and the other planets is a long and sometimes enormously complex one, but a good place to begin is by recognizing that even today nature has not finished its construction project. Some of the same processes first responsible for building and evolving planetary bodies are still going on in our solar system. And the zodiacal light is one of the biggest clues that this is the case.

❧

The dust that creates the zodiacal light is surprisingly fleeting. The smallest grains are so tiny that the ethereal pressure of sunlight, the pitter-patter of photons, together with particle radiation, drives them outward—actually accelerating them to such velocities that they speed away from the solar system and into the depths of space.

But the largest grains are driven inward through a gentle spiraling orbit. Subtle effects from the aberration of sunlight and the glow of their own heat, raised by the warming effect of solar photons, produces a thrust opposing the prevailing wind. Eventually, the increasingly fierce radiation from the approaching Sun can erode or disintegrate these grains, either dissociating them outright into gaseous atoms and ions, or reducing their size enough so that they too get caught in the solar wind and blown back out to the interstellar void.

There's another mechanism that gets rid of interplanetary dust, and that's planetary greed. Over the course of a year Earth's gravity and sticky atmosphere alone capture an astonishing 40,000 metric tons of dust from the solar system. We know this to be true because we can catch that dust. Starting in the 1970s, scientists have used stratospheric balloons and even NASA-operated U-2 spy planes to collect extraterrestrial grains high in Earth's atmosphere. These captured particles have played a direct and vital role in the scientific reconstruction of the history and evolution of our solar system.

As a result of all these ways in which it can be removed from play, interplanetary dust is never long-lived by the standards of planetary physics. Over a period of between a thousand and a hundred thousand years, an average grain will be lost or destroyed. Yet there it all is, glowing peacefully in our night sky. Which means that somehow, from somewhere, the dust is being *replenished*.

It's an important sign that the solar system is not immutable— another piece of evidence, like Tycho Brahe's supernova, that the cosmos ticks to the beat of a different clock, far removed from our human sense of time. This fact affects the way we perceive the universe; it's also going to lead us to a new way of describing the nature of our place among it all.

So where does the solar system's dust come from, and what does it tell us about our deeper history? There are two principal culprits responsible for generating it. One is the relatively benign dispersal of comets, and the other is the violent collision of asteroids.

Glowing comets are produced when small bodies rich in frozen water and other volatile compounds, such as solid carbon dioxide, pass close enough to the Sun to get heated to a critical level. In the vacuum of space, material like frozen water doesn't become liquid when it warms up; it undergoes a direct transformation between a solid phase and vapor. Because of this, the frozen compounds in a comet's solid nucleus can erupt as jets and plumes of gas that spew previously trapped grains of ancient, dusty solids into interplanetary space to form part of the material reflecting the zodiacal light.

We suspect that the rest of this dust is produced by the collision of asteroids. Recently the Hubble Space Telescope has captured just such events happening out between the orbits of Mars and Jupiter. Hefty, boulder-sized objects and lumbering mountains of rock do occasionally hit each other. When it happens, great streaks of material plume outward, trailing these orbits and spreading across space.

In many senses, this means that in zodiacal dust we're seeing the undoing of more than 4.5 billion years of hard work. The sequestered elements and crystallized minerals of planetary origins are being exhumed and dumped unceremoniously to take their chances in the solar wind. Four and a half billion years after the formation of our solar system, the ancient fossil remains are still jostling and grinding, and in the case of comets, evaporating and crumbling. All of these things are like the flotsam and jetsam left after a great storm has passed, partial clues to an ancient past and deep future that are critical components of our effort to understand our cosmic status.

❦

Mapping the temporal, and still changing, nature of our cosmic home is all about setting key events in their correct sequence. But where and when do we actually begin the story of our solar system, and where and when do we end it? I could go as far back as 13.8 billion years, to the emergence of the primordial elements of hydrogen and helium in a rapidly cooling universe that was barely more than three minutes old. Or even earlier than that, to within about a second of the

Big Bang, when a one-in-a-billion deviation from symmetry in the matter and antimatter contents of the universe left a residue of particles that became all the visible matter we know. Another potential starting point is with the earliest stars that began forging heavier elements by fusing the nuclei of hydrogen and helium into oxygen, carbon, and so on.

But to make stars, gravitational forces must first act to sequester material into denser and denser things. It's a process that can involve a condensation of matter by factors of more than a trillion trillion. And of equal importance is the specific history of our Milky Way galaxy, still assembling itself from dark matter and normal matter.

The Sun with all its worlds is like a single raindrop on a particular hour of a particular day in a specific cloud somewhere in the skies of Earth—except that cloud is now mostly gone. So to tell the story of Earth's origin we need to focus on what was once here, a place in the Milky Way galaxy that existed about 5 billion years ago.

The interplanetary dust is a clue to that place. Before it was a part of comets and a part of solid asteroids, some of it was *interstellar* dust—born as scorching stellar plasma, rich in carbon and silicon. This was a gas that chilled down and crystallized as old stars shed it like loose skin, or sometimes expelled it during great supernova explosions. Like windblown sand, these microscopic grains spread out to pollute and enrich interstellar space and form nebulae; the stellar nurseries. These are structures a lot like a place we study today, a remarkable cloud of gas and dust called the Trifid (*try-fid*) Nebula.

From the vantage point of Earth, the Trifid Nebula appears as an interstellar structure resembling a three-lobed flower, some 25 light-years in girth and more than 5,000 light-years away. Within it we can see a slow drama unfolding that echoes our own origins. Although all the nebulae in our galaxy represent only about 5 percent of the stuff between stars, it's in places like these that gas is most densely packed and new stars and planets are forming, as they have done for billions of years.

Massive stars already lurk inside the cosmically thick gathering

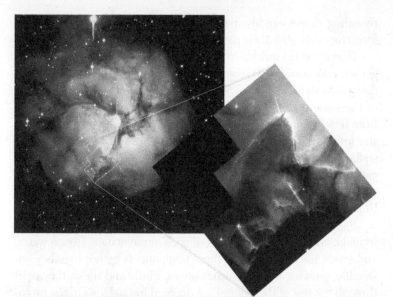

Figure 6: The Trifid Nebula. A detail from a Hubble Space Telescope image, zooming in on one particular region. Nearby stars illuminate the fingers and edges of dense interstellar gas (J. Hester, Arizona State University, and the Space Telescope Science Institute, NASA/ESA)

of gas molecules and dust in the Trifid. A few of these objects are tens of times larger than our Sun, and as a result they are both hotter and more luminous. Their radiation pours out across the Trifid like the wash of a flame across paper. Great interstellar fronts of potent ultraviolet light scorch the cooler gas, dispersing and pushing it into extraordinary sculptural forms. Sharp fingers and ridges of denser gas appear as the thinner material around them evaporates.

The shock-like pressure of this passing flood of light and particles can trigger the condensation of nebula material, collapsing it under its own weight. Gas that would be tenuous to our human senses is pushed over the tipping point where gravity gets a grip and begins to build additional star systems—carving out chunks of fertile nebula many light-years across. As more time passes, this same fierce radiation helps evaporate the unused portions of the gas, revealing dense egg-like patches in which stars like our Sun can form, together with their natal planets.

Gravity works quickly in these structures, gathering material together, concentrating it toward the center. Accelerated matter rains down into these locations. Sometimes this is helped along by more of the same external pressures, including the wash of shock waves from nearby exploding suns. In this central core the beginnings of a star forms, a growing sphere of spinning matter that we call a protostellar object. Squeezed under its own weight, this increasingly hot gas becomes surrounded and fed by spiraling, orbiting material that stretches out in a great disk extending a hundred or even a thousand times farther than the Earth's orbital radius.

Some of this nebular disk of gas cools and condenses enough to make additional dust, grains of carbon molecules, frozen water, and sandy silicates. Some of these components gather together unsteadily, growing to a few inches across. Fluffy and sticky, they swirl through the rest of the gas and are dragged inward toward the protostar. But instead of destroying them this spiraling descent can help them grow further.

As these small clumps sweep through the thickening disk of matter, they gather up more material, with many growing to hundreds of yards across in barely a thousand years. And it's an accelerating process, helped by gravity, turbulence, and the random jostling of matter to compound the growth and produce objects known as "planetesimals." These primitive bodies can grow as large as a hundred or five hundred miles across in anywhere from ten thousand to a million years, depending on where they spend their time. It may not seem like it, but this is a remarkably fast process, going from diffuse nebula to bona fide planet-like objects in the blink of an eye.

Closer in to the developing, and increasingly hot and compact, central protostar, there is a dearth of more volatile solids. Frozen water can't survive in these warmer zones, although water molecules can form a layered gas in the disk. But farther out in this great spread of matter, beyond what is charmingly termed the "snow line," temperatures are low, and frozen water becomes a substantial and vital piece of the building blocks of larger and larger objects. Enormous planetary objects can form in these zones—massive icy spheres that capture the raw nebular gas around them with their powerful gravitational fields, becoming giant worlds like Jupiter and Saturn.

The environment of this great disk is surprisingly good for chemistry as well. Atoms and molecules combine in a bewildering number of arrangements. In fact, long before this nebular material found itself in this situation, it had been doing all manner of chemical experiments on itself. Molecules of water, carbon monoxide and carbon dioxide, and well over 180 distinct compounds have been identified out in the interstellar murk, and they've all been constructed by the basic chemical reactions of raw atoms and ions.

Now, in the dense and circulating matter around a forming star, more chemical processes can really take off. Reactions take place in the gas, in frozen solids, and in the relatively warm and benign environment on the microscopic surfaces of dust grains. And these compounds are mixed and recycled in the turbulent disk, a marvelous mash of chemistry from simple molecules to increasingly complex compounds such as alcohols, sugars, and perhaps even amino acids—the stuff of life.

But the clock is ticking. As all of this action takes place, the disk is simultaneously being evaporated away, puffed back out to interstellar space as radiation from the environment pummels it, the continuation of the erosion we see in the sculpted and scorched clouds of the Trifid Nebula. Once the process of star and planet formation begins, there is a finite amount of time before environmental radiation, including that of the new star at the center, cleans everything out and brings things to a near halt. It's like a field of wildflowers that have only a brief time to grow, bloom, and leave their seeds and stalks before the hot sun dries up their soil and nutrients.

While all of this happens, the star in the center is going through its own significant birth pangs. Spun up by the torrents of matter falling onto it, the baby object can sprout great magnetically governed streams of material that pour away from its poles. These expel around 10 percent of the incoming matter and, most critically, let the protostar rein in and slow its frenzied spinning, which would otherwise prevent it from condensing and shrinking further inward.

The deep insides of the baby star are getting hotter and hotter as it squeezes smaller, nearing the breakthrough point at which sustained nuclear fusion begins in earnest. The first elements to meld are deuterium and hydrogen. This helps stabilize the interior temperature of the protostar, keeping it close to a million degrees Kelvin, throttling the fusion process until it has grown massive enough for full-blown hydrogen-hydrogen fusion to kick in.

It's a wild place toward the protostar's surface. Ultraviolet light is pumping out, and flares and ejections of hot gas erupt to blast the surrounding disk full of condensing gas, dust, and would-be planetary objects. It resembles a giant sputtering engine, getting ready to fire up to full steam. Amazingly, this entire process—from condensing nebula gas to the formation of a brand-new, almost-powered-up star—takes less than a hundred million years. Even briefer is the true protostellar stage from cloud core to protostar, taking barely a hundred thousand years. Compared with the eventual lifespan of the star, this is the equivalent of a seven-hour period in an entire human lifetime.

And in the case of our solar system, at some point during all of this early assembly there was another critical event. It has left us with some of the most important clues that our origins were in a place like this, and it provided something that would help forge our planets into their present states.

❧

The very oldest rocks that we can get our hands on are not native to our continents, or to the Earth. They're meteorites.

This extraterrestrial material comes in a variety of forms. Many are fragments of far larger objects: rich iron and nickel nuggets and rocky boulders that were once deep inside asteroid-sized bodies, embryonic planets that were later shattered and dispersed. This material has been reworked, melted and cooled, and is more similar to the mineralogical forms of the deeper realms of our own world than to anything from the pristine void.

But there are other meteorites that defy any comparison with the terrestrial rocks we're familiar with. These are the true ancients, not processed by any familiar geophysics. They are the most primitive and basic agglomerations of matter we know of, essentially untouched and unaltered since their components formed some 4.57 billion years ago in our condensing protostellar environment.

Ancestral remains like these have been collected in a few places across the globe. Two spectacular examples are known as the Allende and the Murchison meteorites. In what seems to have been an act of sheer cosmic coincidence, both came crashing to Earth in 1969. The first was in February of that year, when it came down in Pueblito de Allende in northern Mexico. Its supersonic atmospheric entry dumped a total of several tons of extraterrestrial material across a 200-square-mile region. The second came in September with a fireball entry over the small town of Murchison in eastern Australia, and left behind about two hundred pounds of primitive matter.

These two meteorites are called carbonaceous chondrites, and they're spooky objects. They have no earthbound equivalent. There is so much carbon and hydrocarbon in these rocks that they are positively oily, tar-like objects, packed with complex molecules that even include types of amino acids—building blocks for biochemistry. When the fresh pieces were collected, eyewitnesses reported that they had a smoky, aromatic smell—probably as some of these chemicals evaporated into the air.

But inside the black matrix are other formations, too: small mineral spherules called chondrules. These are the cooled and solidified remains of molten droplets of rock that were heated and then chilled and frozen in little more than minutes to hours out in space. Later, the drifting and bumping of matter around our baby Sun brought these pieces together, gluing them into larger solids, along with a wealth of carbon-rich grains and dust.

There is another type of component, too—whitish specks, barely a few millimeters across. Consisting of a mix of minerals, these structures are rich in the elements calcium and aluminum, earning them the name of calcium-aluminum inclusions, or CAIs. Their unique properties tell us that they must have formed in even hotter conditions closer in to the proto-Sun, places where they could be heated to well over 1,000 degrees. These too were molten spatters that froze solid into small mineral ashes, only to be collected by us more than 4 billion years later. In these tiny structures is a wealth of vital information about our deep past.

The first astonishing fact about these CAIs is that they are demonstrably older than the Earth itself. Geologists can look at the elemental contents, in particular the mix of lead and uranium, and put a pretty precise date to their formation—somewhere between 4.567 billion and 4.571 billion years ago.

Scientists also find that CAIs have an unexpectedly high content of a particular isotope of magnesium. About 80 percent of the magnesium here on Earth has a total of 24 neutrons and protons in its atomic nuclei. But there are a couple of other stable flavors of magnesium around, with 25 and 26 particles in their atomic nuclei. In the CAIs there is proportionately more of this 26-particle isotope of magnesium than the usual terrestrial ratios. So the question is, what could have been happening 4.6 billion years ago to cause this?

Nuclear physics tells us that the most plausible way for nature to produce an excess of this type of magnesium is for a radioactive isotope of aluminum called aluminum-26 to "decay"—spitting out surplus energy and transmuting itself into magnesium-26. The half-life of this change is about 710,000 years, and we also know that aluminum-26 is produced in abundance when stars go supernova. So, putting two and two together, we arrive at the following scenario.

A little more than 4.56 billion years ago, just before the CAIs stuck themselves together somewhere inside our protoplanetary system, a massive star must have exploded close enough to ram radioactive aluminum-26 into our local environment. This could have happened as nearby as just a couple of light-years from us. Yes, there might be other ways to produce this radioactive pollution, as our system sweeps through galactic matter, but the supernova route is likely the most efficient. So this microscopic clue in meteorites points to a decidedly tumultuous presolar environment. It also provides a natural explanation for something else. We know that when baby planets and rocks collide and crush together, the energy of these violent processes heats them up. But across the solar system, from the interior of the Earth to the iron-nickel meteorites that were once parts of a different planetary interior, dense rocky material has again and again been melted and kept molten far more efficiently than could result from a simple collision.

What could have kept things hot? Well, a spray of supernova-produced radioactive aluminum at the right time would provide ample energy to melt any significant collection of rock. As the aluminum nuclei decay, they dump out energy. Trap this inside anything large enough and the temperature will build to a couple thousand degrees—enough to melt all known minerals.

This heating would have been ferocious. Because of the relatively brief lifespan of radioactive aluminum, we can tell that the contribution it made to keeping objects molten inside was likely at least five times greater at the dawn of the solar system than it is today. And there was almost certainly help from other quarters—other parts of the nuclear mix.

There is now good meteoritic evidence that there was also radioactive iron-60 in the young solar system. This is another product of a nearby supernova, and it turns into nickel-60 during its half-life of about 2.6 million years. In fact, there are nearly twenty so-called extinct radionuclides that show up in meteoritic material as their "daughter" isotopes and point to an array of processes that once made our solar system a significantly more radioactive place.

Many of these isotopes are related to the broader evolution of our galaxy—a steady mix of elements that get sequestered out of interstellar space as a solar system forms. But others, like the unstable isotopes of aluminum, iron, and also calcium and manganese, are locally produced, artisanal products. They were cooked up in a brief period just before the clump of nebula that made us detached itself into a dense little nub of matter—an event that could also have been triggered by the same supernova explosion generating these hot nuclides.

Forcibly injected into our forming system by a supernova shock wave, the total mass of these freshly synthesized, barely million-year-old radioactive nuclei (along with more benign elements) amounts to about 0.01 percent of the mass of today's Sun. That may not sound like much, but that's about thirty-three times the mass of the Earth, sprayed into the material forming the great protoplanetary disk of our young environment.

Together these elements would have ensured that any rocky body bigger than about twenty miles across would melt inside. Eventually, after about three million years, the heat of the radioisotopes would dissipate, and objects would start to cool and re-crystallize from the outside in—with big planet-size bodies cooling most slowly. It seems, therefore, that we're standing atop a true smoking gun—a planet whose most fundamental geophysics was laid down by a wash of radioactive pigments, and whose planetary neighbors shared the same set of circumstances. It's an astonishing connection to the past.

But what becomes of our seething nest of stars, and the explosive siblings that create our radiological history? What happens to something like the Trifid Nebula as millions and billions of years roll by? All of the direct evidence for our original close-knit stellar nursery and its massive supernovas is long gone. It's possible, of course, that over the following millions and billions of years these stellar sisters have simply dispersed, drifting out along great orbital paths around the galaxy and tugged in varying directions by ever-present galactic gravitational tides. But it's also possible that our original home still exists as a great collection of stars that has left us behind.

Astronomers have looked for this stellar Eden, searching for star clusters in our galaxy whose members come close to matching the Sun's elemental composition and age. It's a tremendous challenge. Just figuring out which stars could have shared our original region of the galaxy is limited by our precision in measuring such enormous distances and the specific motions of stars, as well as by the sheer number of objects to sieve through.

One candidate is a collective known as Messier 67, a gathering of suns and stellar remnants some 2,700 light-years away from us. It contains more than a hundred stars that bear a striking resemblance to the Sun. There is a catch, though: recent computer simulations of the motions of stars in Messier 67 have examined the projected path that our solar system would have had to take if it were thrown clear of this birthplace—and it doesn't seem very likely. It would require a very rare alignment of no less than two or three massive stars in Messier 67 to provide the gravitational ejector seat to throw us out to where we are today. And in the course of this expulsion, gravitational tides would likely have ripped our nascent planetary system apart.

However, this conclusion itself rests on assumptions about the configuration of the Milky Way's great spiral arms of stellar objects. If these change more than we thought over billions of years, it's possible that Messier 67 could have let the Sun go in a less dramatic, and more plausible, fashion.

So the jury is still out on where our solar system originated, but the radioisotope clues and the events unfolding in other nebulae leave us in little doubt that, one way or the other, we have been orphaned. Which brings us back to the rest of the story of what happens *within* our forming solar system.

❧

In the great disk of gas and dust around the proto-Sun, it takes only a few million years of agglomeration and collision to form lots of large objects. Out in the colder reaches, beyond what will eventually be the orbits of the asteroid belt, frozen water is stable, and this extra bulk of solid material can combine with rock to build giant icy planetary cores. These massive spheres are some ten to fifteen times larger than the Earth, and their strong gravity siphons up surrounding gas to form huge atmospheric sheaths.

As I mentioned before, the planet Jupiter is one such body, enveloped by an astonishing cloak of matter. Most of this is ancient hydrogen and helium that amounts to more than three hundred times the mass of the Earth. The sheer weight of that matter places the interior of a planet under enormous compression. Even hydrogen is forced into forms unfamiliar to us, such as a metallic liquid state. And with or without the heat of radioisotopes, a young gas giant planet can glow with the thermal energy produced by such compressive formation. Even today, four and a half billion years later, Jupiter still leaks that primordial heat—and its core remains close to 50,000 degrees Fahrenheit.

Toward the inner reaches of the budding solar system—in what will eventually be the orbital zones of the planets from Mercury through Mars to the asteroid belt —there are dozens upon dozens of rocky bodies called planetary embryos, the surviving champions formed out of planetesimal collisions and mergers.

Each contains a few percent of the mass of the Earth, and each experiences a somewhat hazardous existence over the next few tens of millions of years. Individually they don't grow much more, but sporadically they merge together in great crushing collisions that

Figure 7: The Earth and Jupiter to scale. At 317 times the mass of Earth, the great Jove belongs to an entirely different planetary class.

remelt and re-form their mineral contents. Over time a few come to dominate, forming the inner planets.

Beyond the final orbit of Mars there are plenty of embryos, but this is difficult terrain for making planets. The gravitational pulls of Jupiter and Saturn sweep through this zone in ways that can accelerate these smaller objects and make their collisions more destructive than constructive. Perturbed by gravitational pulls, they can be flung into new orbits. Some fly inward, finding their way to join with the inner worlds. Others are spread out elsewhere in the system.

Although some details are obscured to us now, we know that many more things happen over the next few tens of millions of years. Planets experience a certain amount of what's called orbital migration—something we'll return to in detail—as well as the occasional large impact with other bodies. The Earth itself seems to have suffered a massive collision with an embryo about 4.53 billion years ago that resulted in the formation of the Moon.

Our planet was also on the receiving end of a later "veneer" of asteroid impacts. In that explosive peppering, a great amount of the precious substance we call water was deposited onto the young surface, barely cooled enough to harbor such a volatile compound. Here, too, came much of the rich chemical mix that comprises the outer layers of this planet, often recycled into the upper reaches of our molten interior, but also vital in setting in motion the chemical machinery of the surface, oceans, and atmosphere.

Other planets are different. Venus appears to have retained an earlier outer layer of rocky material. Unlike the Earth's, this stratum was never peeled off by a Moon-producing collision. Some theories also suggest that Venus formed as two huge planetary embryos collided almost head-on—a process that could explain its strange east-to-west spin that only completes one revolution after more than an entire orbit around the Sun.

Mars is smaller, a tenth of the mass of the final Earth, and its bulk composition is somewhat different. Proportionally more of the volatile elements made their way into Martian rocks. But it too experienced enormous collisions with planetary embryos. This is a possible reason for its strange global geography—a shocking north-south disparity, with thin planetary crust and smooth plains across the northern third of the planet and thicker crust and rocky highlands over most of the southern hemisphere.

Intriguingly, Mars and Venus may have also harbored more temperate, Earth-like climates back in these early days 4 billion years ago. Both are long gone, Venus's replaced with its thick carbon-dioxide-rich atmosphere and high surface pressure, resulting in broiling temperatures of over 800 degrees Fahrenheit, and Mars's replaced with a dismally thin and dry carbon-dioxide-dominated atmosphere. This tenuous air is barely 0.6 percent of Earth's atmospheric pressure at the Martian surface, and temperatures veer with season and location from minus 200 degrees to over 70 degrees Fahrenheit. But Mars holds out the most hope of another viable environment for life, with clear evidence for liquid water having once flowed and accumulated on its surface, and a mineralogical and chemical state not unlike many places on Earth.

These planetary atmospheres are fickle and leaky things. Gravity is what holds an atmosphere like the Earth's as a thin blanket around us. But the atoms or molecules of gases are in constant motion, and the higher the temperature, the higher the average speed of these constituent particles. Outliers can reach escape velocity, sailing away to space. These escapees tend to consist of the lightest components, and for this reason Earth has long since lost any atmosphere of hydrogen or helium that it once harbored. Today, if atmospheric water molecules are broken apart by ultraviolet light or particle radiation, the hydrogen atoms can find their way up and out of the Earth's clutches.

Our planetary magnetic field helps limit this loss by protecting our upper atmosphere from the full ferocity of stellar radiation. It's a good thing it does, because every lost hydrogen atom is lost for good, along with the water molecule it once belonged to. A planet can literally dry out this way, and this kind of mechanism may well have helped transform Mars from a much wetter and warmer place to its present desiccated environment.

Earth, too, is not what it started out as. The surface environment has evolved and varied enormously in temperature and chemistry across the eons. The most ancient minerals, crystals of zircon, provide evidence that liquid water has almost always been present somewhere on or close to the planetary surface. And, critically, for about the first 1.5 billion years after Earth's formation, the atmosphere held very little of the reactive element oxygen.

This changed, and it changed because of a truly extraordinary phenomenon on this planet—life. Sometime around two and a half billion years ago, organisms such as the single-celled cyanobacteria got the upper hand in their ecosystems and proliferated. Their metabolic tools produced an abundance of oxygen, whose rising concentration transformed the planet over the next billion years.

Other characteristics have varied, too. Past temperatures on Earth have on average sat well above today's levels, by several degrees. But they also seem to have occasionally plummeted to levels that would almost encase the planet in ice. There are also deeply rooted chemical and geophysical cycles that tend to push our climate toward what might be considered an uneasy equilibrium—maintaining liquid water on the surface as the atmospheric contents regulate the loss of thermal energy.

Threaded deeply into this near-bewildering network of planetary mechanisms are living things. At any given time there are trillions upon trillions of them—blooming and dying, eating and decaying, and constantly reworking the world. But in cosmic terms these are all such pitifully small details, generally inconsequential changes to planetary characteristics, much like the gentle weathering of long-exposed fossils. Indeed, the bigger picture speaks to a rather different perspective on our existence here than our parochial human outlook usually gives us.

❧

That greater vantage point is one of the critical pieces that we need in order to tease apart the tangle of Copernican mediocrity and its counterarguments and to begin to formulate answers to the puzzle of our cosmic significance. So let's imagine for a while that we are an external observer of our galaxy, the Milky Way. With an omnipotent gaze we are able to watch this complex gathering of more than 200 billion stars, great volumes of gas, dust, and dark matter, as it evolves across not just centuries or millennia, but through billions of years. We also have a particular fondness for ordinary stellar objects, and one of those is the Sun.

When we first come across it, this lone beast has just ignited its central core with the blazing fire of hydrogen nuclear fusion. The energy from this furnace emerges in two main ways. One is a persistent flood of subatomic particles known as neutrinos. These ghostly little things barely interact with anything else, and even the dense bulk of the Sun is largely transparent to them as they streak away and out into the universe at close to the speed of light. The other component of fusion energy is a thick flux of photons that diffuse their way out through 400,000 miles of solar plasma, eventually emerging into space as a glow of visible, ultraviolet, and infrared light.

This rich wash of radiation warms the planets, asteroids, comets, dust, and gases that orbit the Sun. For the inner worlds it dominates their surface environments, pumping energy into their circulating atmospheres, and even into an ocean of liquid water on the third planet out. But as we track this little star it slowly changes. In its first 4 billion years it becomes more luminous by about 30 percent, powering a diverse sprawl of living things on the third planet throughout that time. After approximately 10 billion years it is twice as bright as it was in its youth. With tender regret we can recognize the signs of aging, the inevitable progress toward eventual death.

Unlike many other phenomena in the universe, stars like the Sun become more and more brilliant as they grow old—at least for a while. As the single protons of hydrogen nuclei fuse together in the stellar core to make nuclei of helium, they alter the fundamental composition of the star, enriching it with this heavier element. As a result the interior becomes denser and hotter, and the rate at which hydrogen is consumed gradually increases (think of a bonfire that slowly collapses in on itself, and burns increasingly hot and bright).

For the wet planet encircling the Sun this has a profound effect: even by 6 billion years that growing luminosity has pushed its surface climates to extremes that can no longer comfortably sustain liquid water oceans. But at 10 billion years that is the least of the problems for this world and its immediate neighbors. As the Sun finishes off the last drops of hydrogen in its core, it begins its difficult and painful transition to the stellar afterlife.

Over a period lasting a little more than a billion years in this distant future, our star becomes more and more bloated and disturbed. Its outer regions swell in fits and starts, eventually growing so large that they engulf the innermost worlds, almost reaching the orbit of the previously wet planet as a looming, reddened ball of plasma. At the same time, this once pristine star sheds enormous quantities of its material, blowing gas and rapidly condensing dust out to interstellar space. It may ultimately lose almost half of its mass this way. This loss profoundly alters the gravitational dynamics of the planets around it, whose orbits adjust by also expanding—following the rules deduced by a single small sentient being called Isaac Newton more than five billion years earlier.

What drives these dramatic outward changes in the Sun is a series of internal rearrangements and processes. Once the central hydrogen has been used up, the stellar core begins to shrink and rise in temperature. It leaves just a thin shell of fusing hydrogen around it, a bit like the flickering outer rim of a fire that's recently burned out. Eventually, though, this shrinking core becomes hot enough for helium fusion to ignite. It's a process that requires temperatures of 100 million degrees, ten times higher than those for hydrogen fusion. This next reaction is also less efficient, but it turns helium into two new elements—carbon and oxygen. Over the next hundred million years, the increasingly dense stellar core and its flow of energy cause the outer part of the star to grow even further, until the helium fuel runs out, too.

It's a pivotal moment for our orphan star. After approximately 12 billion years, and fewer than sixty orbits around the Milky Way galaxy, it has finished consuming all it can. It is not massive enough to ever raise its central temperature to the levels needed to fuse carbon nuclei together, and so there is no new source of energy, nothing edible left in the larder.

In short order the stellar engine comes to a halt, its final flares of energy acting to swell and push away the last vestiges of its outer layers, blowing them off to interstellar space and creating a beautiful nebula spread across tens of light-years. Eventually, only the inner core of the Sun is left behind, mostly naked and exposed. It's made of carbon and oxygen, held up against its own weight by strange and fundamental forces due to the quantum nature of the submicroscopic—where the dual wave-particle behavior of matter creates a resistance to gravity's compression.

We refer to this bizarre object as a white dwarf. It has no energy source. It is simply a hot ember that will take trillions of years to cool down. And as it does so, its constituent atoms arrange themselves into a lattice, a regular array—it crystallizes. The Sun's distant future is to end as a giant and darkening carbon-oxygen jewel in space.

Gazing at this speck we see that some of its original planets have survived. In fact, what was once the third world out has narrowly escaped destruction during the stellar death throes. It now orbits almost twice as far away from the system center than it first did, since the Sun has lost some 40 percent of its original mass. Frigid and barren, this world endlessly and hopelessly circles the increasingly dim white dwarf, the only thing that is left of its mother.

Thus ends the ten-billion-year spree of this one star that we decided to take an interest in. But we have no time to mourn, because there are already new ones just like it to pick and choose from. While we have watched our favorite star live its glory days, another ten billion or so suns have been born throughout the Milky Way.

❧

Our solar system's birth was a flurry of physical and chemical activity, most of which took place in barely more than a few tens of millions of years. Afterward, billions of years pass, in what amounts to a relatively benign fossil state during the lifetime of a single modest star. But from our human perspective this takes an eternity, and is filled with a tapestry of complex action.

Living things existed billions of years before us, and in the briefest of flashes we emerged from a web of astrophysical, geophysical, and molecular evolution. In one microscopic sliver of that brief flash I manage to stand upright on a Chilean mountaintop to contemplate my place in the universe. Before me, the folds, wrinkles, and contents of the landscape are a direct consequence of the molten geophysics of the Earth—their deeper origins in the radioactive elements forged in massive stars in a long-lost and unimaginably ancient stellar birthplace.

It is a convoluted trail that leads to, and passes beyond, this fleeting moment. Even though the underlying rules may be simple, the pathway through the cosmos to you or me is riddled with twists and turns. That's important, because one of the potential ways in which we might learn about our cosmic significance or insignificance is by asking just how many roads lead to life like us, or, for that matter, life at all. To chart that, our next step has to be to ask about the stories of other planets—other worlds around other suns across this galaxy and those beyond it. What they have to tell us is amazing.

3

NEIGHBORS

In the quest to understand our place in the universe, few objects have grabbed as much attention as exoplanets—the cosmic oases that we've long hoped might be out there. With good reason, too, because it's obvious that a scarcity of other planets, especially other Earths, could severely alter our outlook. A few worlds scattered to distant and inaccessible corners of existence would make the search for life impossibly hard.

The idea of other worlds, places beyond "here," isn't just deeply rooted in science. As we've seen, it's been at the metaphorical center of various schools of philosophical thought; it's also surfaced time and time again in human art and literature.

A good example comes from an ancient source, the marvelous tales of *One Thousand and One Nights*. These clever stories were gathered and compiled from generations of storytelling and folklore more than eleven hundred years ago, and they're still terrific entertainment.

One of my favorites is the adventure of a young sultan called Bulukiya who goes on a quest for the herb of immortality. This mission takes him to places full of bizarre and unearthly things, from clusters of heads and birds growing out of tree limbs to the recursive depths of hell.

In these travels he also encounters a cosmic angel and gets a quick lesson on the status quo. This magical being tells him that there are no fewer than forty worlds beyond the Earth, each forty times bigger than Earth, and each home to exotic creatures beyond his wildest dreams. It's a wonderfully entertaining fantasy. It also makes it abundantly clear that the inventive minds of humanity's storytellers have long conceived of many worlds beyond ours—worlds alien enough to make a mere mortal collapse in awe.

What might lie beneath, above, and far outside our normal existence continues to be great fodder for human imaginations, from C. S. Lewis's allegorical Narnia to the bustling universe of *Star Wars*. Sometimes, though, we can lose track of our most inspired creations until nature surprises us by bringing them to life again. We've recently found ourselves in precisely this situation—not with cosmic angels or quests for immortality, but with planets beyond our own solar system.

The surprise is not just that there are other worlds, but that they have qualities that challenge our imaginations—lifting us out of a hole of mundane thinking. As I'll show you next, this reality brings to light one of the biggest clues in our quest, a critical piece of the puzzle of our significance. But its impact is not simple, because on the one hand what we've discovered strongly reinforces the Copernican outlook (that we are mediocre, not central), and on the other it provides some of the best evidence to date that there is something unusual—perhaps even special—about our circumstances.

❧

Finding planets around other stars is extremely difficult. There is no other way to describe it. The reasons are straightforward enough: planets are small and dim, while stars are big and luminous. And stars and their planets appear incredibly close together when viewed across interstellar distances—a problem, because light's properties make even a perfectly constructed telescope blur its images. Brilliant starlight in these systems swamps the feeble glow of planets.

Of course, most of us have seen the powerful glare of the full Moon in our sky, and even spotted the bright dots of planets like

Venus and Jupiter. The planets we know don't seem so shy. But this local experience is very misleading.

A giant world like Jupiter reflects sunlight and also emits a gentle infrared glow from its deep, warm interior. But altogether, the maximum amount of electromagnetic energy coming from the brightest solar-system planet is only about a *billionth* of that pouring out from our Sun. A planet like the Earth, warmer but much smaller than Jupiter, is an equally dismal squib. While we think the Moon is bright, that's essentially an illusion of circumstance. The lunar surface actually reflects barely 10 percent of the sunlight hitting it—about the same as a lump of coal. It appears bright to us only because it's so close-by, and because the Sun's light is still strong at our location.

If we were viewing our solar system from light-years away, planets like Jupiter and the Earth would vanish from sight, drowned out by the glare of blurred starlight—like motes of black dust next to a brilliant camera flash. To see these worlds directly requires huge telescopes and clever optical trickery—technology that is only now creeping up over our horizon. But there are other ways to penetrate the dazzling cloak of stellar systems to try to sense the presence of planets.

One approach I've mentioned before, and it goes back to Isaac Newton. He pointed out that stars are orbiting the center of mass, or balance point, of a system. With no planets present, that center lies at the center of the star, but with planets in a system their gravity offsets it. That position itself isn't usually constant, because as the planets whiz around their orbits to different locations, the fulcrum point must also move.

In other words, if planets are present, stars wobble around, and that wobble changes with time. You might be able to spot it directly, as the star moves back and forth by tiny, tiny amounts across the sky. But you might have slightly better luck finding it by using the Doppler effect: the telltale shifting of the frequency, or color, of light as the star moves away from us and toward us.

It's still a ridiculously difficult measurement to make. A planet like the Earth induces a motion of the Sun that is little more than a few *inches* a second, and that motion only reveals its signature pendulum-like cycle over the period of a year. Jupiter might be a better target. It can shift the Sun by about fourteen yards a second, but the pattern of that swing is spread across the ten years of Jupiter's orbit. You'd still have to be very patient and very consistent in your observations to spot it.

In case that isn't tricky enough, a star's surface is also a turbulent place, with upwelling and down-welling motions of its scorching and luminous gas. These localized motions can readily exceed the steadier en masse movements due to planetary pulls, adding confusing and confounding signals to the stellar light we see.

The quest is not for the fainthearted. Starlight captured by powerful telescopes must be split out into thousands of its component frequencies, like the rainbow formed with a glass prism. Astronomers must extract the delicate and specific spectral features of electrons jumping back and forth within stellar atoms to serve as a yardstick. These markers need to be measured with exquisite precision, monitored, and carefully converted into an estimate of the speed of a thousand-trillion-trillion-ton object that may be moving more slowly than a walking human.

There are other ways of finding planets, often equally hard because they rely on serendipity as much as skill. Sometimes planetary systems are oriented so that here on Earth we can catch worlds eclipsing their parent stars, blocking a few fractions of a percent of the star's light from reaching us. Spot this, and spot it again on the next orbit, and the next, and you have a clue to the presence of those dark specks, and even their size.

More rare, and more complex to interpret, is the signature that stars and their planets make by distorting space and time around themselves, bending the path of light by their gravitational fields—a consequence of the relativistic nature of the universe. When the light from a more distant star passes through just the right place in an intervening stellar system, it's as if a lens were being held up in space. That light is briefly magnified into a flash that glows for a few days before the swirl of stellar motions through our galaxy takes things out of alignment again. A single star can produce this gravitational lensing, but add some planets around it and the character of the flash can be altered in ways that reveal something about those worlds, their orbits and their masses.

Difficulties abound with all approaches, and the longer history of attempts to find planets around other stars is littered with failures and unsubstantiated claims. However, by the latter half of the twentieth century these astronomical techniques had progressed to the point where a bold and persistent cadre of scientists figured they had a realistic shot at detecting the tiny dark specks of worlds around other stars. That is, if they existed—something that seemed likely, but was still surrounded by nagging doubts.

Interestingly, most of these scientists thought that if they found anything, it would be some pretty dull stuff. They essentially imagined replicas of our own solar system, familiar types of planets in familiar types of configurations. Although contemporary science-fiction writers had kept wide-eyed inventions like those of the Arabian Nights alive with even more outrageous-sounding speculations, these were not the worlds explicitly sought by researchers. None of the hypothetical planets and orbits that astronomers imagined were extraordinary—they were all close facsimiles of our own immediate surroundings.

In part it was respectable scientific conservatism that kept other notions at bay. And it was also an adherence to Copernican principles that temporarily misled us. If we weren't central or special, then it stood to reason that other places would look like us. If we were just an average planetary system around an average type of star, we'd expect other planetary systems to be similar.

This meant that by the late twentieth century we were effectively looking for planets like Jupiter and Saturn. These would be massive worlds in large, slow orbits, producing a very languid but conceivably detectable dance in the motion of their stellar parents. Finding other planets the size of Earth was still well beyond the sensitivity of these early experiments, although there is no doubt that these smaller worlds lurked at the back of everyone's mind as an ultimate goal.

Our solar system was also the sole template for theories of planet formation. Scientists' thinking about the origins of planets from the gas and dust of interstellar space had varied across the centuries. But by the latter half of the 1900s there was a general consensus on the mechanism. As I've described, there are very well supported physical reasons why planets can form out of a great disk of gas and dust surrounding the shrinking and gathering nebular matter that builds a star. And the solar system has a very specific arrangement of smaller, rock-rich planets forming closer to the hot Sun, while larger, gas- and ice-rich worlds form farther away. This arrangement was, and still is, the poster child for ideas of how worlds form.

It was hard for people to think outside this box. There is even a numerically appealing, empirical rule called the Titius-Bode law, which originated in the 1700s, when astronomers puzzled over the arrangement of our solar system. This law predicts the distances of planets from the Sun using nothing more than a simple algebraic series, a sequence of numbers. Specifically, the sequence is 0, 3, 6, 12, 24, 48, 96, 192—each number after 3 being double the number before it. The "magic" formula was to add 4 to each number and then divide by 10 to get the mean distance of each planet from the Sun in terms of astronomical units, a measure of Earth's distance from the Sun. The positions the formula yields are not exact, but they're close. This pattern certainly hints at a deeper principle at play, a way in which planets would be formed and arranged that could be universal. That is . . . until you take a closer look.

Over time, scientists have realized that the Titius-Bode "law" is at best just the tendency of natural phenomena to blindly follow mathematical forms. These are functions called power laws or exponential curves. At worst, the pattern is just complete coincidence. It's a rule that applies in our solar system, but not necessarily anywhere else. An idea like this has a powerful grip, though, and while the reasoning might not have been articulated at the time, this appearance of a "law" surely contributed to the general feeling among scientists that all planetary systems should look like ours.

Looking back at this now, I'm a little shocked. It's as if our species were so traumatized by the adoption of the Copernican Principle that we could do little more than shuffle along with our heads held low. Having correctly displaced Earth from the center of things, most astronomers took this averageness as a new gospel. It was hard to allow for the possibility that these "unexceptional" circumstances were merely one instance born out of a countless number of configurations and pathways.

One could therefore say that it was an act of cosmic justice that the first irrefutable proof of worlds beyond our solar system came in the way it did: in the form of something so different that it was immediately clear we had managed to blind ourselves to what was possible. Planets, it turns out, may be the ultimate nonconformists.

❋

Ten miles inland from the northern coast of the Caribbean island territory of Puerto Rico is a terrain of thick, ancient, and slithering jungle. A lot of this flourishing plant and animal life sits on top of porous and soluble limestone, and in some places the wet millennia have eroded the rock away to create huge sinkholes and slumping hollows of land. Usually these are lush places, cups of moist fertility nestled between gentle mountains. Except for one spot.

Here in a great bowl a thousand feet across, instead of trees and undergrowth, the land is tented over by more than 38,000 tightly fitted, sieve-like aluminum sheets—as if a metal seal has been carefully placed to wall off the humid ground. Looming five hundred feet above this silvery surface is an equally massive framework. Three towers at the bowl's perimeter attach to thick steel rigging that hangs in the air to intersect above the center. At this middle point a complex mesh of cables and girders holds a bulging structure formed by an intricately assembled tessellation of triangular plates, a vital part of a sophisticated listening post for extraterrestrial radio waves.

It's a remarkably immodest beast of ultramodern technology, and an outrageous thing to find in what is otherwise a tranquil and somewhat remote corner of paradise. Its name is the Arecibo

Observatory, and while it sits quietly among the trees, there is nothing at all shy about it or what it does.

In February 1990 this colossal telescope was listening to the faint stirrings of electromagnetic radiation coming from a distant patch of our galaxy—a place 12 thousand trillion miles away, a journey of two thousand years at light speed.

These electromagnetic fields bounced off the aluminum plates resting in Arecibo's giant bowl and converged onto its sensitive detectors, suspended in the air. Although dimmed by its long interstellar trip, this radiation had its source in the ferociously spinning remains of the interior of a sun that had died some 800 million years earlier.

This object is a neutron star, a remnant consisting of the nuclear particles called neutrons together with a smattering of the ones called protons, as well as some electrons. It's all that is left of a star somewhat bigger than the Sun that once burned brightly, until the nuclear reactions in its core could go on no more. When the power switched off inside it, the center collapsed under its own weight. It was a catastrophe that produced a huge supernova explosion, blasting the outsides of the star to space and leaving behind this fearsome sphere.

Unlike any type of matter that we have experience of here on Earth, the stuff of a neutron star is packed together very, very tightly. There are no atoms or molecules, just what is in effect one giant nuclear ball glued together by gravity. A neutron star with a mass twice that of the Sun is only about twelve miles across. Close to its surface the gravitational acceleration is so intense that a fall of just three feet would bring you crashing down at over 1,000 miles per second.

Neutron stars can also spin at an exceedingly fast rate. Since they are born out of the uncontrolled collapse of the insides of a star, there are various ways in which they can get spun up, and some neutron stars spin all the way around in *milliseconds*. Typically they're also very hot, a million degrees or so. They're bristling with energy: magnetic fields and electrically charged protons and electrons protrude and flow from their surfaces. In combination, these properties can create one of the most surreal of all phenomena in the universe—the pulsar.

A pulsar beams electromagnetic radiation into space like an unstoppable lighthouse. Its intense flood of energy reaches across the galaxy in a great spiral of outwardly rushing photons. This radiation front sweeps through tens of thousands of light-years in a tiny fraction of a second as the neutron star spins around.

The dense bulk of an object like this results in huge inertia. So, like an enormous flywheel, it can take eons to dissipate enough energy to slow down. As a result, its rate of spin is stable. In fact, the radio beacon of a rapidly spinning pulsar can maintain a timing precision comparable to that of atomic clocks on Earth.

So it was very surprising when the radio signals reaching Arecibo in early 1990 contained the clicks of a neutron star spinning around 167 times a second—and something else, a mysterious small variation in the timing of the pulses of radiation that defied immediate explanation. One of Nature's clocks was looking a little erratic.

Over the next two years the observatory repeatedly listened to this object. As astronomers pored over the data, they saw that this strange signal variation was cycling through the same pattern again and again over the course of a few months. The only possibility that made sense was that something was tugging at the pulsar, forcing it to move in its own little orbit around an off-center gravitational balance point in the system. Except that balance point wasn't the result of just one other object nearby, but of several, and they weren't huge— they were planet-size.

In January 1992 the astronomers Aleksander Wolszczan and Dale Frail published their discovery in *Nature*. They had done what so many had tried to do. They had found convincing evidence for the first exoplanetary system in the data from this distant pulsar—the first set of other known worlds in our galaxy.

Today, with far more observations made of this system, we know that there are at least *three* planet-size bodies orbiting the pulsar. Two of these are approximately four times the mass of the Earth and orbit some 34 million and 43 million miles from the pulsar, even closer than Mercury's mean distance from our Sun. The third

Figure 8: An artist's impression of planets orbiting the pulsar known as PSR B1257+12 (NASA/JPL-Caltech/R. Hurt [SSC])

one is tiny, barely 2 percent of the mass of the Earth, and comparable to the mass of the Moon. This diminutive planetary scrap orbits within the others.

Facts and figures like these do little to create a useful mental picture, so let me put it another way. This system is so strange, so utterly different from ours, that it instantly confounds any sane extrapolation from what we thought we knew.

Here the planets have no normal star. Instead, all they have is the poisonous remains of one, a fearsome mother that they hug in a tight orbital dance. A spinning pulsar spews harsh and destructive radiation into its environment and heats the planetary surfaces with a steely light. When its ancestral star died a billion years ago, it did so in a mammoth supernova explosion, obliterating any other worlds that might once have existed around it. These odd planets we see are the ghoulish remains of what once was: the detritus from that epoch of destruction reassembled from the dust, coagulating and condensing under gravity's thrall to build these recycled globes, mock worlds that will never bask beneath a normal sun.

Nothing could have prepared us for this. Our first sighting of other planetary bodies revealed a system drawn from the astrophysical underworld. Yet there it was, proof positive that objects like planets existed beyond the confines of our solar system. And this place was just softening us up for the next surprise.

Three years later, in 1995, astronomers announced the discovery of the first robustly detected planet orbiting a normal, Sun-like star, in a system only 50 light-years away from us. This was another pivotal moment in science: finally we had confirmation that other stars like ours could harbor planets—something we might never have doubted, but had no proof of before.

Like the pulsar worlds, this new planet was spotted by its gravitational influence on its parent star, forcing this sun to move in its own little orbit at the balance point between these bodies. It was the same behavior of stars and planets that Isaac Newton had described almost four hundred years earlier, a direct consequence of his theory of gravitation. This stellar wobble could be seen in the shifting frequency of light from the system. But there was a catch.

This new planet completed an entire year's orbit, a full cycle, in just over four Earth days. In fact it was only 5 million miles away from its parent star, far closer even than Mercury at its eccentric orbit's closest pass to our Sun—a safe 28 million miles. Not only that, but instead of being some dwarfish runt of hot rock, this planet was a giant more than a hundred and fifty times the mass of the Earth.

It's safe to say that no scientist or natural philosopher, during more than two thousand years of recorded thought on the matter, had ever spent much time considering the possibility that a system like this could exist. Indeed, the theory of planet formation had gotten to a point where there were good reasons to think that there was no way a giant planet could possibly exist so close to its parent star.

A big world like this simply couldn't form anywhere but much farther out in a system, where the combination of orbital dynamics and cooler temperatures would allow such a planet to bulk up.

Only a few scientists (such as the astrophysicists Peter Goldreich and Scott Tremaine, who fifteen years earlier had studied how planets might "migrate" inward through a protoplanetary disk) had ever considered ways that planetary objects might end up in unexpected places. So, while this discovery was a triumph for the astronomers who'd doggedly persisted in making these difficult measurements, it was also tremendously puzzling.

Since these first discoveries, the surprises have continued to pile up. We're finding that the diversity of exoplanets, and their apparent refusal to conform to our notions of how planetary systems should be, is just astonishing. We might have expected these other worlds to be a little different, not exactly like ours, but we hadn't and couldn't have imagined just how varied they were. They fill up the landscape of possibilities. They bring some of our central questions of our own planet's cosmic significance into a brand-new light. This bewildering array is our introduction to what is, in effect, comparative planetology—the categorization and classification of species, and the study of what makes them all tick.

❧

So let me welcome you to this league of extraordinary worlds. It's not an exclusive club by any means, because almost anywhere you look you will find its members, but it disturbs our rather pitifully parochial outlook on the universe.

Much of what follows is based on informed extrapolation, and we are already beginning to test and verify many of these speculations with new telescopic observations and clever techniques to tease out signals that reveal planetary size, temperature, and even composition. Let's go in and take a look at the club lounge, resplendent in its livery of finely crafted furniture and regal occupants.

Over in that corner by the fireplace are the giant planets orbiting perilously close to their parent stars. These are representatives of those very first exoplanets discovered around normal stars. By now they have gained an unofficial name: they are the "hot Jupiters" (although there is often only passing resemblance to this familiar giant world).

They shouldn't be where they are, but there's no denying their plump and sometimes rosy presence. Perhaps some of them have migrated to where we find them, pulling and pushing against the great disk of material that used to surround their forming planetary system, squeezing their way to the front of the line. Or perhaps they've been flung to these spots by getting a little too close to the gravitational influence of other worlds, performing an ill-conceived dance move that left them facing the scorching wrath of their suns.

Some of these giant worlds are orbiting so closely that they complete one entire circuit in barely twenty-four Earth hours and are heated on their day sides to temperatures of more than 1,000 degrees. Tidal forces have pulled many of them in so tightly that they no longer have normal days and nights. They are permanently locked so that their day side is *always* their day side, while their night side is perpetually dark, cooling into the chill void of space.

This strange state of affairs has produced some fearsome climates on these giant worlds. The glowing heat of the day side can drive the flow of atmosphere into the night side and around the entire girth of the planet at supersonic speeds—forming great jet streams of crashing, shock-wave-producing atmosphere. With no surface of mountains or continents beneath the gas, it chases itself without a pause.

High temperatures on these worlds result in all manner of chemistry and atmospheric compositions that we can barely recognize from our own solar system. Carbon monoxide, vanadium oxide, and titanium oxide gases are found here, influencing the layers and structures within these planets. Clouds are made not of water or ammonia but of silicates and iron, scorching hot clusters of heavy atoms. Hardly the stuff of imagined fluffy animals on a summer's day; more like nightmares of time spent in Hades.

And the hot Jupiters are not above cajoling their stellar parents, either. Their gravitational tug can raise tides and waves within a star's own atmosphere, and their powerful magnetic fields may interact directly with those emanating from the star itself. No longer immune to its surroundings, a star can be affected by its planets, more than the other way around. The solar atmosphere sporadically flares and brightens with a hint of petulance as the bulk of a hot Jupiter buzzes around like a persistent insect.

But lest you think that these planets are all smug giants sitting in their chairs close to the roaring fire, consider that some of them are doomed to die. They can overstay their welcome. Gravitational tides can gradually erode their orbits, causing them to spiral inward over time spans of a few tens of millions of years. In time they either plunge beneath the stellar surface or are shredded by tides into a short-lived ring of debris around the star.

Some of these giant worlds bring this fate upon themselves for even more perverse reasons. While all the planets in our own solar system orbit in the same sense that the Sun spins—clockwise, if you like—roughly one in five hot Jupiters does the opposite. These renegades orbit against the spin sense of their parent stars, a retrograde motion. This places them in the perilous situation of having their orbits inevitably eroded until they spiral to a horrible fate.

This unfortunate orbital direction is confounding. As far as we know, the first stages in the formation of stars and their planets set them spinning and orbiting in the same direction. Anything else would precipitate rapid dynamical disaster, with planets trying to move counter to their own protoplanetary disks, and such worlds are unlikely to form. So where do the retrograde exoplanetary objects come from?

As with so many members of this league, we simply don't know for sure. But it is possible that these planets formed at far greater distances from their parent stars, in proper "prograde" motion, only to be flung into extremely noncircular, elliptical orbits by gravitational jostling with other planets. These orbits could end up pointing right out of the plane of a system, from where they will literally flip over to become retrograde, just like a spinning hoop falling on one face or the other. Eventually, gravitational tidal forces from the star circularize their orbits, bringing them in close, where we find them.

These varied life experiences have left the hot Jupiters with a rather interesting array of characteristics. Some of them have become exceptionally bloated, their diameters inflated well above expectations, resulting in their overall density becoming very low. A few of these giant worlds have an average density less than that of liquid water. Others have undergone a range of chemical and structural changes due to their proximity to a stellar energy source and the history of their formation.

This is most marked in their outward appearances, the upper layers of their atmospheres. In these intense environments the dominant compounds are almost unrecognizable compared with the soothingly cool, but nasally caustic, wisps of crystallized ammonia and methane we might find in our own Jupiter or Saturn. In the most extreme cases, temperatures are high enough that even iron atoms can play a role like water does in cooler places, forming a cycle of high clouds and condensation into heavy metallic raindrops.

Some hot Jupiters have atmospheres that are exceptionally rich in carbon, offering a clue that their deeper interior may also be chock-full of carbon to an extent unfamiliar to us. These giant planets may give rise to huge diamond layers in their cores, and they hint at the possibility that other, more modestly sized worlds may exist that are also made more of carbon than of silicon, a plausible but unfamiliar scenario.

The substances like gaseous titanium and vanadium oxides that exist in these conditions can contribute to an atmospheric surface that sometimes absorbs almost all light falling onto it. There are worlds that are more radiation-absorbent than the darkest coal or charcoal. Pitch-black planets. Except that the light flooding them is so bright, so intense, that the human eye would still see its reflected glow—like an imperfect chameleon trying to blend into the inky darkness of the cosmos.

Hot Jupiters are truly a class unto themselves. But sitting off to one side of them is another group, a select bunch of daredevils and hot-Jupiter wannabes. These, for want of a more official name, I'll call "Icarus worlds." Unlike the hot Jupiters, these planets have large orbits, taking months to complete a single cycle. Their orbits are not circular—in fact, they are at the opposite extreme of orbital shape: narrow ellipses with one end sitting tens of millions of miles away from the parent star, and the other end dipping down to within shouting distance of the stellar furnaces.

For some of these Icarus worlds, the change in stellar heating they experience during an orbit is eight-hundred-fold. At their slowest-moving far points, conditions are temperate. But as they drop inward and zoom through their nearest pass to the star, their temperature climbs by about 700 degrees in the course of just a few hours.

Every time they come close to their parent star, gravitational tides sap a little more of their momentum. Over millions and millions of years they will give up these ridiculous orbits—most likely the result of playing gravitational bumper cars with other planets earlier on—and gradually conform more and more to paths like those of the hot Jupiters. Eventually, they may even join that group, sidling over to the big armchairs by the hearth, but ultimately they may be consigning themselves to doom in the stellar fires.

It isn't just giant planets that hover precariously close to stars; there are smaller planets of rock and metal compositions that parade themselves within a few tens of millions of miles of their stellar parents. Some of these, a few times more massive than the Earth and likely denser, have surfaces heated to temperatures well beyond the melting point of all conceivable rock types.

Without the protective envelope of a giant's atmosphere, the outer layers of these planets become an ocean of lava, a perpetual hell. Even metallic compounds like aluminum oxide can boil off the surface, only to recondense as dusty particles that may in turn be blown clear by stellar winds in a great plume of pollution, like the exhaust from a cosmic smelter.

It is possible that these worlds were once more like Neptune in our own solar system, a planet covered with a thick blanket of primordial gas and ices. Perhaps a process of migration brought them into their present orbits, where their protective sheaths were eroded and evaporated. It is also possible that they were always a simple body of rock and metal, just unfortunate enough to end up getting pushed into their present awful conditions.

•

So in this end of the exoplanetary lounge, nestled up by the fire, are a wide range of worlds. But just steps away are an even greater diversity of objects and startlingly unfamiliar systems. Here, for example, in another set of chairs, is a group of nine major planets surrounding a single star.

It may at first not seem so strange—after all, we have eight major planets around the Sun, plus numerous trans-Neptunian bodies like Pluto. Nine is hardly impressive. Well, no, it wouldn't be, except that these nine planets are orbiting their star (which happens to be almost the same mass and age as the Sun) at distances that would place them *all* within the orbit of Jupiter.

With the exception of two of these worlds that are only slightly more massive than Earth, the rest are big and chunky: ten, twenty, even sixty times the mass of our little home. And even though they're all crammed into what might seem to be an awfully tight-fitting system, there's still space for some more. In places like this it is as if the processes of planet formation simply ran unchecked, churning out planet after planet and somehow dodging the nefarious effects of gravitational interactions between them. You feel like going up to these systems and saying "Well done, well done!"

By now something should be apparent: planetary systems, and planets themselves, show enormous diversity. This diversity is fascinating in itself, but it also raises serious questions about how we quantify our cosmic mediocrity, our commonness. We're not the only planetary system anymore, but to add insult to injury, so many of these new worlds seem to buck all our expectations of normalcy.

In some systems it's the orbits of the planets that are special. Gravitational dynamics have evolved the motion of these objects into arrangements where the periods of the orbits, the planetary years, are synchronized according to simple number ratios. For example, an inner world may make two orbits in precisely the same time that an outer planet makes one orbit. It is as if the motions are forming part of a well-tuned musical instrument, shifting pitch in perfect harmony.

This phenomenon is known as resonance. Here the orbital motions have locked onto this rhythm because the planets repeatedly arrive at the same points in space after the same interval of time. Because of this, their gravitational forces consistently tug at each other the same way, maintaining their synchronization. During the formation and history of these systems the planetary orbits must have slowly evolved and become trapped in this state, snared by their mutual gravitational attraction and unable to escape.

Although numerous examples of this type of orbital resonance exist in our own solar system, they are almost entirely confined to the motions of minor planets and moons; there is no resonance of motion among our large planets that quite matches these exoplanetary systems. For example, little minor planet Pluto and giant Neptune share a resonance in their orbits such that for every two Plutonian years, there are three Neptunian ones. Several moons around the giant planets are locked into specific patterns. Around Jupiter the moons Io, Europa, and Ganymede follow a 4, 2, and 1 pattern in the number of orbits completed in a given interval. None of our major planets exhibit this behavior with each other—at least not anymore, since there is evidence that once, perhaps 4 billion years ago, Jupiter and Saturn shared a one-two tango.

These resonances happen surprisingly often across our galaxy, given that they're a moderately special situation. But there is another characteristic of the orbits of many planets that is absolutely critical to discuss, because it is both extremely common and extremely different from the way things work in our own solar system.

We have discovered that the majority of the planetary league orbit not in circles, but in gracefully elliptical paths. These are the same elliptical forms that Johannes Kepler discovered as the solution to the misbehaving patterns of motion in our own solar system, and the same forms that arise naturally from Isaac Newton's gravitational laws. But the Earth is in only a very mildly elliptical orbit, deviating from a pure circle by only a couple of percentage points. In fact, no major planet in our system is more than 10 percent away from a circle except for Mercury, which deviates by 20 percent from a circular orbit.

By contrast, surveying the planetary league, we find that 80 percent of all exoplanets are in orbits of *more* than 10 percent ellipticity.

In fact, more than 25 percent of planets anywhere follow orbital paths that are *50 percent* more elliptical than a circle. In other words, if we try to find the solar system's seat in the planetary league, we would have to look quite hard for the few spots left for the likes of us. With such relatively circular yet large orbits, our solar system is in the bottom quarter of the table of elliptical motions. It's distinctly unusual.

The elliptical flavor of orbital architectures points toward a number of critically important things. For one, it suggests that the majority of planetary systems, perhaps over 70 percent, have gone through episodes of what's termed dynamical activity. This means that in the past the planets were likely positioned differently, passing closer to each other at times and pulling harder on each other through gravity. Over time this can cause a certain amount of change and even disruption; planets get flung around, sometimes lost or relocated. I'll return to this remarkable characteristic a little later on when we tackle the evolution of planetary orbits and how they relate to the Copernican picture of mediocrity, but it indicates a far more tumultuous history in most systems than anything experienced in ours.

The other aspect of elliptical orbits that is important to our goal of evaluating our cosmic status has to do with climate. Many possible cousins to the Earth are typically subjected to far more dramatic shifts in the amount of energy they receive from their parent stars over the course of their year. That energy is a critical factor in determining the surface environment of such planets, and so it is of the utmost importance.

Diversity in the league of planets doesn't stop here; orbits are merely one of many distinguishing characteristics. Many, many systems contain numerous examples of another class of planets that are not represented around the Sun at all. These are worlds that range in size from slightly more massive than Earth to five or ten times more massive.

They are the super-Earths, the smallest of which have at least a passing resemblance to our own planet—although they may not be truly "Earth-like," a quality I'll get to a little later. In fact, the larger

variants may be very different. Many appear to have huge atmospheres, possibly containing lots of hydrogen. Some of these bulky
objects are likely covered in vast quantities of water. They may be
frozen solid. They may also be awash in a global ocean that reaches
to almost unimaginable depths—tens, even hundreds of miles—with
pressures and temperatures such that the physical and chemical behavior of water becomes unlike anything we experience on Earth.

Others may have a modest splash of water, or none at all. But
many should be persistently volcanic. Despite having disruptive surface environments, these places are also fertile in terms of chemistry.
The continual conveyor belt of upwelling hot rock constantly re-
freshes their chemical mix, impregnating their landmasses with a
rich stew of highly reactive compounds. The large size of these planets also means that their geophysical lifetimes are very long, their
cooling surfaces smaller in proportion to their volumes. Billions of
years' worth of pent-up activity will keep them looking youthful far
longer than their more petite, Earth-size cousins.

This area of the planetary league, the middle of the lounge,
seats a huge population. Super-Earths, together with slightly larger
Neptune-like worlds and smaller, Earth-size objects, are so abundant that they have to sit almost on top of each other to fit in. And
our present surveys indicate that their preferred configuration is to
be in closely packed orbits that may take only days or weeks to complete a single circuit. It actually seems that this might be the default
type of planet formation across the Milky Way. In fact, the data suggest these bodies could readily outnumber the count of stars in the
Milky Way . . . there may be hundreds of billions of these planets.

And here is another surprise, another wrenching twist away
from our preconceptions about our ordinary Sun and our solar system that calls into doubt some of the Copernican precepts: most
of these planets orbit around stars that are smaller and dimmer than
the Sun, because most stars in the universe are smaller and dimmer
than the Sun.

Take a census across the galaxy and you'll find that 75 percent of
all stars are less than half the mass of the Sun and less than a few
percent of its luminosity. The smallest, about one-tenth of a solar

mass, are barely one-thousandth as luminous. They're faint reddish balls of hydrogen and helium littered across the cosmos.

Most of our immediate stellar neighbors are such objects. Within twenty light-years of us there are 8 stars like the Sun or slightly larger, but there are 101 known stars that are smaller. Even the famous Alpha Centauri system really consists of three stars. Two are more like the Sun, but one—Proxima Centauri—is barely 13 percent the mass of the Sun and less than 0.2 percent as luminous.

All such stars are so dim that no naked human eye has ever seen one of them; they become visible only through the light-gathering optics of a telescope. But before you dismiss them as underlings, clouds of little gnats in interstellar space, consider this: not only should they harbor the majority of planets in the galaxy, but these small stars are also the most long-lived of all stellar bodies.

Lower internal temperatures, combined with a turbulent nuclear digestive system that recycles material, result in these stars taking an enormous length of time to exhaust their hydrogen fuel. And they do it with incredible completeness. Over about ten billion years of steady nuclear fusion, a star like the Sun will never consume more than about 8 percent of its hydrogen before veering off into rapid decline. By contrast, a much smaller star may manage to consume 98 percent of its hydrogen and take more than a *trillion* years to do so.

This means that in looking around at the league of extraordinary planets, you will find that the overwhelming majority are in dim stellar systems that steadily pump out energy for their small rocky and icy offspring for a hundred times longer than we can expect the Sun to. I think it's reasonable to speculate that an external observer of the Milky Way, armed with astronomical instruments, would survey our realm and quickly conclude that this is the pecking order for planet-hosting stars: small ones rule the roost, and larger ones are much more of a rarity.

And now, at the chilly far end of the planetary lounge is a heavily shadowed set of chairs. But these appear to be as full as any others.

Plopped in the murky depths of their seats are some of the most mysterious members of the club—the interstellar worlds, the rogues, the free floaters. These planets have *no* star to orbit. They are adrift in open space.

Occasionally they're revealed by the effect they have on the passing light of distant stars. The lens-like distortion their masses make in the fabric of space-time briefly magnifies and diverts these rays around their otherwise cold and dark frames. Perhaps orphaned by the brutal gravitational pulls in some young planetary systems, these objects have been ejected, flung out from their stellar nests to wander the galaxy.

There is evidence for a remarkable number of these wandering planets—possibly as many as there are stars in the Milky Way. Their existence profoundly alters the balance of astrophysical objects in the cosmos, shifting it away from giant structures toward the small condensations of planetary matter produced in the turbulent circulation around starbirth. Again, a diversity and population size we had not anticipated.

Taken all together, the lounge in this club holds an astonishing array of members, and as we look around we keep noticing more and more types. The truth is that I've barely scratched the surface by focusing on the species that at present we know the most about.

For example, there are also planets in lots of systems with more than one star. Talk about different! Twin suns, sometimes even more. Often these are places where the stars orbit the system's center at a considerable distance. In such cases planets can safely form and orbit around one or the other star without being unduly disturbed by the gravitational pull of its luminous sibling. But there are also places where planets orbit right around twin suns, two stars at the center of the planetary system. These orbs rise and set together in such distant skies, sometimes eclipsing each other and sitting side by side all in the course of a single day.

Astronomers are now finding that there must also be a rainbow of possible planetary compositions and circumstances. Worlds blanketed with atmospheres of water vapor or molecular hydrogen, ocean planets with no continents at all, carbon worlds of novel geo-physics, and icy snowballs plunged into permanent winters so deep that even their atmospheres have frozen and fallen to the ground.

There must be hot, cold, warm, and lukewarm worlds, and sometimes all these zones in one. Some worlds are young. Some are ancient. There will be chemically rich worlds, and some of these will be awash in unfamiliar compounds, while some are more like the Earth. There will also be chemically poor ones. There will be worlds with rings of dust or ice, like Saturn. There will be worlds surrounded by moons; some of those moons may be as large as Mars or the Earth, perhaps even carrying their own atmospheres, oceans, and landmasses.

Several things are clear as we survey all of this. The first is that neither our star nor our planetary configuration may be representative of the most common environments in which small, rocky, wet planets are found. In other words, even among all this diversity, the Earth and its circumstances are somewhat unusual.

This is intriguing. Let's suppose that life-nurturing environments are equally likely around any stellar type or in any orbital architecture. If this is the case, we might expect the great majority of habitable worlds to exist around low-mass stars and in elliptical, or closely packed, orbital configurations. They or their companions should be super-Earth worlds. Thus, purely on this basis, we might expect to exist in one of those systems, *not* in the kind of system we live in.

There are a number of explanations. One is that it is just chance that we don't live in one of the most common types of life-bearing systems; we're a bit of an outlying probability. If this is correct, there's nothing profound to learn; we just happen to live in a some-what atypical place. This could mean, for example, that life thrives in all manner of places that seem quite alien to us, from planets or-biting faint low-mass stars to even more exotic locales like icy or temperate moons around giant worlds. If life happens often, it's more likely for there to be instances of it in less-common places like our solar system.

But another possibility is that life-nurturing environments are *not* equally likely among all stellar types and orbital architectures. Perhaps there really is something that makes our circumstances especially suited for life. This latter option would mean that the universe as a whole might produce a smaller amount of life than it would otherwise. If you recall, I mentioned the question "How common is life in the cosmos?" as a puzzle without a clear answer from either Copernican or anthropic thinking. If this second scenario were to prove true, it would offer us the beginnings of a way to measure life's frequency—and the likelihood of abiogenesis (the natural origin of life from nonliving matter), a critical subject I'll come back to.

There are some obvious planetary properties that might contribute to making a system more or less likely to harbor life. Temperature is a key one. Earth exists in a slightly uneasy equilibrium that sustains large amounts of liquid water on and near its surface. Liquid water is an extraordinary natural solvent that plays a central role in terrestrial biochemistry and in the geophysical behavior of our planet. The precise distance of the Earth from the Sun, the present-day luminosity of the Sun, and the atmospheric composition of the Earth all play major roles in keeping us awash in oceans and precipitation.

But we still don't understand all the mechanisms at play in maintaining a temperate climate on a planet. My colleagues and I have, for example, investigated how planetary orbits, axial tilts, and even day lengths could alter the climate across planets similar to Earth. It's not very straightforward. Planets on orbits far more elliptical than that of the Earth can still keep liquid-water environments, while planets with shorter days transport less heat from their tropical equators to their poles and may be more susceptible to "freezing out" in ice ages to end all ice ages.

The list of positives and negatives goes on. There are also wholly different watery environments, like those we think might exist beneath the icy crust of moons such as Europa, Ganymede, or geyser-spouting Enceladus in our own solar system. Lakes, or even oceans, of liquid water may exist underground in these places, with no reliance on the warmth of a star.

We clearly need more information to know how to rank these

possibilities, and in the coming chapters I'm going to sift through the facts to see what else can be learned in order to help resolve the issue of what makes a place amenable to life. But the diversity of exoplanets may have something else to tell us in our quest for life. From the way their orbits are arranged to the properties of composition and structure of the planets themselves, it's a menagerie. But as much as this great range of qualities teaches us about planetary astrophysics, it also creates some substantial hurdles for science.

Understanding the mechanics behind planet formation and evolution gets a lot harder when the overall phenomenon clearly has many intertwined parts. It also creates an obstacle that is directly relevant to our search for cosmic significance: If this diversity means that effectively no two planets are ever going to be equal, how do we evaluate our place among these worlds?

To rephrase this: scientists are often keen to talk about the quest for "another Earth," or for "Earth-like" planets. It's an easy way to try to encapsulate the search for worlds that resemble ours in some key aspects—from size to composition to, of course, surface environment. But hidden in these innocent phrases is a whole lot of pain.

The term "Earth-like" comes across as meaning another planet that anyone could recognize, complete with continents, oceans, clouds, forests, and small furry animals. It means that our world is the template, the master copy with which to compare anything else. This contains a subtle shade of the old arguments that simply assumed life elsewhere was a lot like life here.

Actually, what I think we are *really* looking for are the Earth-*equivalent* planets. Equivalence is in this case the same as when a car-rental company tells you that no, you can't have the red, sports-trim, soft-top car you had booked, but you can have an equivalent one, which is neither red, sporty, nor remotely open to the elements, but does come with four wheels and an engine.

At its simplest, the requirements of Earth equivalence must include a surface environment similar to that on parts of Earth today, or during parts of Earth's history. That means temperatures commensurate with liquid water, the presence of water, and chemical fuel and raw materials. It probably also includes a degree of stability, without too many violent changes happening in quick succession or too much biologically destructive radiation.

The fascinating question is whether or not this type of Earth equivalence can be found in places that are outwardly quite different from our home planet, and we'll have to wait and see about that. However, before we move on from the league of extraordinary planets, there is one other lesson they have to teach us, one that is often overlooked. You may think that the implications of planetary plethora I'm about to describe are self-evident, but there are subtle, and important, consequences.

❧

In the years since Arecibo's first discovery of planetary objects beyond the solar system, we've detected thousands of worlds around thousands of suns. We know that these numbers will continue to increase, because already we have enough data to make statistical extrapolations, to estimate our galaxy's total population of planets, to rough out a census. Many scientists have done just this, and the overall pattern is clear.

If we consider just those worlds similar in size to the Earth— let's say, in a range from half its diameter to about four times its diameter—it is obvious that there must be anywhere from a few billion to a few tens of billions of these planets in the Milky Way. In fact, if we consider only those orbiting their stars at the right distances to allow for moderate surface temperatures and liquid water, some studies suggest a galaxy-wide population of more than *20 billion* and even as many as *40 billion*.

With such an abundance of worlds, there is a 95 percent chance that one of these temperate worlds exists within 16 light-years of our Sun, a cosmic stone's throw away. With today's telescopic powers, such a place would be on the cusp of more detailed study. With tomorrow's generation of telescopes and instruments we could hope to seek the signs of life.

To state the fact of this planetary wealth is simple enough, but it fundamentally alters the nature of our questions about life elsewhere. Suppose that the Earth was the only planet in the universe. We might still want to ask what the probability was of life arising on a world like this, but it would effectively be impossible to answer this question. As tempting as it would be to think that the probability must be very high (or else why would the single planet in the cosmos have life?), there would be no way to verify this with only one example to go on.

A second planet in this hypothetical universe would change all that. Whether it harbored life or not, it would allow us to make a mathematical statement about the probability of life occurring on planets, together with an estimate of our uncertainty. More planets would improve the situation, each yes or no question helping us define the frequency with which life pops up on any planet.

So here's the important yet subtle thing. We know that we live in a universe dripping with planets. That means that we live in a universe where the question of life's probability, the chances of abiogenesis on a suitable world, *can be answered*, given enough time and technological skill.

It's not clear that the cosmos had to be this way. Planets could have been scarce and we might still be here, on a lonely Earth, asking the very same question, but forever bereft of an answer. The discovery of the sheer wealth of planets circles right back to the idea I talked about at the start of this book, the Anthropic Principle. One could argue that not only does the universe appear to be tuned to allow at least one occurrence of life; it also appears to be tuned to allow life to find out about itself, to determine the probability of abiogenesis.

We don't know exactly what conclusions we can draw from this—at least not yet. But it is certainly provocative, and it is a fact that we'll need to fold back into our thinking as we explore further, not just in space, but also in time.

Coming to grips with the universe of planets has forced us to think well outside of our local box. We've had to revisit many of our ancient fantasies about worlds unknown. We've also had to correct ourselves, and stop thinking about our solar system as a good representative of the whole.

If the technical challenges of detecting even the nearest exoplanets weren't so great, we'd have reached this point much sooner, and so the surprises are going to keep coming as we struggle to look deeper at these dim specks around their glaring stars. As much as the abundance of planets confirms our Copernican thinking, their diversity has muddied the waters. There are signs that we inhabit a somewhat unusual place, and there is a hint of an expansion to the notion of cosmic fine-tuning. However, the story isn't over yet.

That's because this league of extraordinary worlds represents just a snapshot in the history of our cosmic neighbors. Any comparisons we make with our own solar system are also often based on a simple set of measurements fixed in time. Circumstances today represent a single instant out of 4.5 billion years of past and another 5 billion years of future for the Sun and its worlds. Does it make sense to base all our conclusions on such a narrow view? It would if planetary systems were clockwork—endless, unchanging, and predictable. But they're not. So in the next chapter I'm going to reveal one of the dirtiest secrets in celestial mechanics, because it explains why we really must include the passage of time and the possibility of change in our equation for significance.

4

A GRAND ILLUSION

The year was 1889, and Henri Poincaré at age thirty-four was riding high. A young husband, new father, and rising professor at the University of Paris, he had recently been elected to the enormously prestigious French Academy of Sciences. Just months earlier, in the summer of 1888, he had confidently submitted what would turn out to be the winning entry to a grand-prize-giving competition: an apparent answer to one of the most persistent and challenging problems in all of mathematical physics. Life was looking pretty good.

It may seem a little strange now (although the tradition does still continue for some outstanding problems), but in the late 1800s it was quite common for important unanswered mathematical problems to be incorporated into contests. This case was somewhat special, though—the event's patron was King Oscar II, royal head of Sweden and Norway. Not only had Oscar studied mathematics at Uppsala University, he maintained a close relationship with the academic world. In fact, he had a special interest in the newly established journal *Acta Mathematica*, which was being published by what would later become the University of Stockholm.

It was only a matter of time before someone had the bright idea to hold a royally sponsored competition, whose winning entries would be published in the journal. And so, in 1885, the announcement went out, and a jury of elite mathematicians from Europe and America was selected. The contest asked for any answers to four outstanding mathematical questions chosen by the jurors, although entrants could also select their own topic. For an extra flourish, the prize would be announced to coincide with Oscar II's sixtieth birthday, in early 1889.

The first question, at the top of the list, was extremely well-known and perennial. It was called, simply, the "n-body problem." This problem had a significant history, going all the way back to the late seventeenth century, when Isaac Newton formulated his laws of motion and gravity. Newton's rules explained the forms of planetary orbits brilliantly, and at a superficial glance it would seem that you should be able to apply them to calculate the future motion of *any* set of objects engaged in mutual gravitational interaction.

It could be three objects, or four, or an arbitrary number n. After all, each body would pull at every other body in a way that was readily determined from Newton's law of gravity. So if you knew the starting points, you could surely work out all subsequent motions to any degree of precision you desired.

This was relatively easy to do for two objects like the Sun and a single planet, but Newton quickly realized that it was a whole other story for any more elaborate system. It clearly irked the great Isaac that he couldn't see a way to solve the equations, and he wrote, "to consider simultaneously all these causes of motion and to define these motions by exact laws admitting of easy calculation exceeds, if I am not mistaken, the force of any human mind."

He was, in his typical fashion, pretty much correct. It's true that no few lines of algebra and no straightforward application of integral calculus can yield a mathematical curve that traces the motions of n bodies interacting gravitationally. However, despite the master's proclamation, the n-body problem remained an unresolved and nagging issue. A proper mathematical proof was needed, and perhaps, just perhaps, a more mathematically sophisticated approach could be used to find a solution.

In the time between Newton and Poincaré there had in fact been some good progress in finding more closely approximate ways to map out the orbital evolution of planets. At the end of the 1700s the scientists Pierre-Simon Laplace and Joseph-Louis Lagrange had

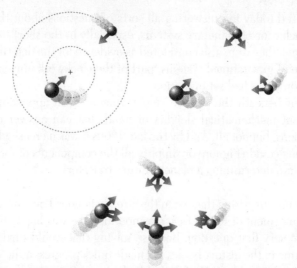

Figure 9: An illustration of the rapidly increasing complexity of objects interacting via gravity. *Top left*: two bodies pull at each other and orbit, a steady and calculable situation. But three bodies (*top right*) involve 3 three-dimensional spatial coordinates, 3 three-dimensional velocity vectors, and 6 three-dimensional force vectors. Four bodies (*bottom*) involve 4 coordinates, 4 velocity vectors, and 12 force vectors—all in three dimensions and all acting simultaneously. No wonder Newton gave up on trying to find an algebraic solution.

each come up with a set of mathematical tools that could at least predict the broad-brush behavior of multiple planetary motions over thousands, and even a few millions, of years. Part of the key came from a very technical insight. Both scientists realized that orbits in a system of multiple objects are "quasiperiodic": the influence of planets on planets means that each one won't always complete its orbit in precisely the same amount of time. It's possible to actually exploit that property with some mathematical trickery to predict the general trends in orbital evolution in a system.

The biggest drawback of these methods is that they don't track every instant of a system's motions; they essentially compute the average of the ways in which planets will pull at each other, or *perturb* each other, cycle after cycle. They are clever techniques, and are still used today for answering all sorts of questions about the general behavior of planetary systems, especially in the short term. At the time, they were also considered as evidence for the deterministic nature of gravitational systems, part of the clockwork universe that Newton's laws had set in motion.

But beneath the surface veneer they are still approximations, brilliant mathematical sleights of hand that can answer certain questions, but not all. And by the late 1800s it was increasingly clear that one couldn't ignore or simplify all the components of force that went into determining a planet's future trajectory.

So it's not surprising that when the already famous Poincaré saw the announcement of King Oscar's competition, he was happy to settle on the very first question, because solving that would further seal his name in the history books. He made quick progress as he worked away at the problem. He felt that he had a mathematical proof that one could determine the stability of a system of *three* gravitating bodies. And, most important, he claimed he could also calculate their motions to any necessary precision. It looked wonderful, and although it solved the problem only for three bodies, it was enough to convince the jury—and Poincaré's prize money was in the bag.

This was where his headaches would really begin. As promised, his winning entry headed off for publication in the *Acta Mathematica*. But as the paper was being copyedited, Poincaré began to realize that something was amiss—he had made an awful error. His proof of a solution to the three-body problem was incorrect, it didn't work, and he had to tell the journal editors. He had overlooked a subtle possibility for the geometrical behavior of mathematical functions that were key to his proof.

Unfortunately, by the time he told the editors, they had already printed the paper and were sending it out across the world. In an effort to contain the disaster, they recalled all the copies, and Poincaré had to foot the bill, which was substantially more than the sizable prize he had received from King Oscar just a short time before. Poor Poincaré. Seldom had a mathematical error cost anyone so much.

There was a silver lining—although not to Poincaré's bank balance. As he recovered from embarrassment and worked through his mistakes, he came up with what would become an enormously influential analysis. It stated that there could *never* be a straightforward answer to the *n*-body problem. In the language of calculus, there was no analytically integrable solution to the general problem of *three* bodies interacting via gravity, and by implication, not for any higher number of bodies, either.

According to Poincaré, if you had a star with two planets orbiting it, there was no way that you could compute the *precise* future (or past) behavior of that system with paper and pen. With more than two planets, any arbitrary *n*-body system, the task would be even more hopeless. The only possible exceptions were a few highly contrived special cases where, for example, the third body is too small to exert any significant gravitational force.

This was quite a thing to say, and Poincaré's new mathematical approach hinted at a side of the universe that would only begin to fully emerge from behind the thick drapes of classical physics in the next century. This quality of the cosmos, as I'll soon discuss, is *chaos*.

It turns out that Poincaré had made tremendous progress when he asserted that the general problem of *n*-bodies was unsolvable, but it would be discovered that the details were even stranger. Getting closer to the root of the problem was not easy at all, and it took almost a century for the next stage of the answer to be produced. In the 1990s some beautiful work by a Chinese mathematician named Quidong (Don) Wang demonstrated that the *n*-body problem could indeed have a complete algebraic solution. There was a hitch, however, and it was a really big hitch: the solution involves the summation of a series of mathematical terms that number in the *millions* of components. In other words, you actually could write down an algebraic equation to tell you the gravitational behavior of *n* bodies— but it might take you forever to do so. And by the time you added up all the pieces, you would have introduced enough rounding errors to make the answers worthless.

•

Here is a critical hint to the true underlying nature of planetary systems, a nature that has become increasingly apparent since Poincaré. The equations describing them exhibit an inability to contain and control tiny computational uncertainties, small numerical errors that eventually rear up and take a big bite out of your ability to predict anything. Nature itself is also full of real variations, and the web of interactions in a planetary system can make it extremely sensitive to these changes. A grain of microscopic dust here or there can change the eventual trajectories of entire worlds, given enough time.

This sensitivity of a system, and of the equations describing a system, is a fundamental property in nature. It is often called nonlinearity, since there is no simple one-to-one correspondence between any changes made to a system and the way it responds. It's a bit like cautiously poking a large dog with a stick: the same small prod may elicit either a timid yelp or a furious and justified attack—the response is nonlinear. And nonlinear systems are special, because they can behave in a way that's chaotic.

Strictly speaking, this is not the chaos of demons and devils and the abandonment of reason and all order, but a mathematical type of chaos—a chaos that may or may not lead to disorder and destruction (contingent on the finest details). At its core is unpredictability, the impossibility of knowing what the future holds. So that grain of dust, or that variation in the lumpiness of a planet, or its altered position in an orbit, may not only result in a dramatically different future path; that future path cannot always be predicted. The same is true of many other complex systems. Nonlinearity applies to the climate and weather on Earth, and to the vagaries of economic systems and stock markets. Uncertainty is built in at the deepest levels.

This type of chaos can be well entrenched in planetary systems, and the fact is that all planetary systems have the potential to be chaotic. This is a real double whammy for the n-body problem and the calculation of orbital trajectories over any long period; you can't solve the equations of motion in any practical way by hand, and even if you could, the system might veer off into an unpredictable chaotic state. This is a truly inconvenient truth that Poincaré had the dubious pleasure of touching upon.

❧

Luckily, in the century after Poincaré's groundbreaking work, a new tool would come along to allow us to explore these jungles of dynamical possibilities. That tool is the computer: thin wafers of chemically altered silicon from some long-dead star, once assembled as part of our planetary geology and then mined, chemically purified, and recrystallized by humans to build microscopic machines to shuffle electrons back and forth.

The beauty of the computer is that by sheer brute force of number-crunching power, we can try to directly model the behavior of gravitational systems. We can simulate the pull between planets at any instant and their resulting trajectories—second after second, week after week, year after year, and eon after eon. In fact, what we can do is apply the tools of calculus—the mathematics of infinitesimal pieces—to build virtual worlds, virtual planetary systems that behave almost identically to real ones, chaos and all.

And the real twist with these computer models is that not only can we simulate a billion years of planetary motion in mere hours or days, we can do it again and again to follow as many of the unpredictable futures as we can stomach. While chaos may reign, we can at least begin to understand how many of the possible futures lead us in certain directions, and therefore build a map of likelihoods— the probability of one type of outcome over another.

The explorers of this virtual landscape have made many spectacular discoveries. Some of the pioneering computer experiments on the long-term nature of planetary motions in our solar system were carried out by Jacques Laskar, then at the Bureau des Longitudes in Paris, and by Gerald Sussman and Jack Wisdom, both at the Massachusetts Institute of Technology, in the late 1980s and early 1990s. Using a variety of mathematical approaches, these scientists attempted to track changes in orbits that could take place over millions, even hundreds of millions, of years as minute shifts in conditions were propagated forward. The researchers even investigated the properties of the past solar system, reversing time and unwinding orbital history, with Laskar pushing back as far as 200 million years into what might have been our dynamical heritage.

At this time, other gravitational experimenters had already studied the behavior of subsets of planets, either the inner worlds or just the outer giants, Jupiter and its cohorts, and even the vagaries of Pluto's lonely orbit. But now the entire system of major planets was set in motion, and the remarkable results confirmed a long-held suspicion. The solar system itself feels the potent touch of chaos.

Over a period of just a few million years, the motions of the planets show what is called exponential divergence. In other words, after this amount of time, the most-impossible-to-measure variations in positions and speeds end up taking the planetary orbits onto trajectories that could not have been predicted. That's not to say that they need be crazy changes, but just that we cannot know what they will be with any real accuracy.

It's a bit like sending out a flock of homing pigeons. If you release them directly from their roost, they simply fly around for a few minutes before swooping back to gobble up their feed, and you can pretty easily track them. You may even know when their graceful flight paths are going to bring them back to the roost—a particular turn and pattern of behavior that is familiar and predictable.

But if you take them off to some remote field before you release them, it becomes much harder to accurately predict when they'll all make it back home. If they're well-acclimatized birds, they will zero in on their familiar roost. But the vagaries of geography, air currents, and the nature of pigeon brains will make that journey very hard to map out in advance.

While we might not be overly shocked by the slightly unpredictable behavior of pigeons, the unpredictable behavior of planetary motions in our own solar system is enough to cause us nightmares. This is a *profoundly* disturbing discovery. Newton's physics and its application by scientists like Laplace had appeared to be describing a clockwork universe, a reality based on laws that could always lead you from point A to point B, through space and time. And although the concepts of chaos and nonlinearity were well-known by the time these numerical computer experiments were carried out on planetary motions, this was the first real confirmation that our solar system was neither clockwork nor predictable.

Our narrow human lifetimes, and even the length of time our entire species has staggered around the surface of the Earth, make us witnesses to nothing more than the thinnest slice of the orbital history of our planetary neighbors. The unending variation in their motions might not seem so ominous or frightening were we omnipotent beings a billion years old, but to the short-lived scraps of biomolecular assemblages that we are, it comes as an awful shock to realize that we're just riding one wave crest in a choppy ocean of orbital possibilities.

But apart from upsetting our perceptions of the reliability of the planetary foundations of our very existence, is there anything more that this unnerving property tells us about the nature of the solar system, and for that matter, about any other solar system? Indeed there is, because in this case chaos can most definitely lead to destruction.

You might be wondering how we can possibly predict the behavior of a system that I've just said is inherently unpredictable beyond a few million years' time. That's an excellent question. The best way to think about this is to consider each possible future configuration as one of an endless set of trajectories, a bit like the path taken each time I throw a ball down a field.

If I could somehow map out every three-dimensional track traced by the ball, and I threw the ball a thousand times, you'd end up with a great bundle of wiry-looking lines drawn in space. Most of those lines may be close together, but there may be a few off to one side that lead to the ball bouncing more erratically, springing off some unseen mound of turf before heading into the undergrowth. If I study just those outlying trajectories and ask what happens to the ball next, after its first bounce, I can selectively choose possible futures for the ball that may lead to something more dramatic.

It's the same way with the future trajectories of the dynamics of a planetary system. After a few million years we can pick those outcomes where the planetary orbits appear to be more extreme, more

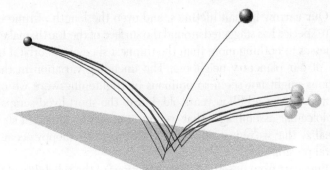

Figure 10: Throwing a ball down a field. Most bounces take it in the same direction, but every so often there's an outlier that takes the ball into the bushes.

likely to cause trouble by bringing objects closer to each other, not farther away. Perhaps it's an increase in the orbital ellipticity, altering the near and far points of an orbit. Or perhaps it's in the orientation of the ellipses, again bringing objects into greater proximity.

We can harvest these futures, and then see what happens to variants of them over the next few million years, before repeating the process and pruning away the less interesting ones. We still can't predict a specific outcome beyond four or five million years, but we can ask what *could* happen, and, to some extent, discover how likely or unlikely certain futures are.

Two scientists who have pursued this kind of question are Konstanin Batygin and Greg Laughlin at the University of California, Santa Cruz. Using computer simulations of the gravitational interactions of planets, they have experimented with the far distant future of our solar system, pushing more than 20 billion years ahead in time, even beyond the death of the Sun.

It turns out that we need not look so far down the line for interesting things to happen. While the planets of the outer solar system—Jupiter, Saturn, Uranus, and Neptune—have a good chance of remaining in stable orbits for well beyond the next several billion years, the planets of the inner solar system are a different story.

In one possible future trajectory, the planet Mercury will fall into the Sun in approximately 1.26 billion years—its orbit having been perturbed and thrown into disarray by interactions with other worlds. In another case, in some 862 million years, Mercury and Venus may collide. Even before this occurs, Mercury's errant ways will result in Mars being flung clear of the solar system, ejected to wander interstellar space for eternity.

The Earth's orbital future in all such cases will of course be altered as well, pushed into new configurations—and, more likely than not, into utter disaster. These experiments, together with other key findings by Laskar and his colleagues, reveal a number of unappealing possibilities for us. In a few billion years, previously distant worlds like Venus and Mars may become our nemesis, colliding with the Earth in events that could only be described as the end of everything as we know it.

Are these outcomes at all likely? Predictability is of course the core problem, but we can certainly evaluate just how many future trajectories out of the multitude could end up in these severe cases. For Mercury's orbit to evolve into something more elliptical, and more vulnerable, than its present configuration, the odds are anywhere

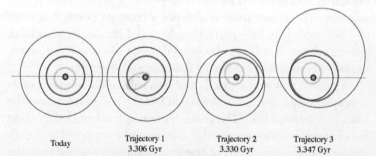

| Today | Trajectory 1 3.306 Gyr | Trajectory 2 3.330 Gyr | Trajectory 3 3.347 Gyr |

Figure 11: Possible futures. *Left*, our present-day solar system, the orbits of Mercury, Venus, Earth, and Mars. *Right*, what could take place in around 3.3 billion years (Gyr) with a probability of 1 percent. Mercury's orbit can become distorted enough to collide with Venus (Trajectory 1); Mars's orbit could intersect with that of the Earth (Trajectory 2); destabilization could result in Venus and Earth colliding (Trajectory 3).

between 1 in 100 and 1 in 50 over the course of the next few billion years. This may not seem like too much to worry about, and clearly our species may well not be around to experience it, but tucked into these simple odds is a big shift in our conceptualization of celestial mechanics. Indeed, there's far less mechanistic predictability and instead a stark and unnerving mathematical likelihood that our solar system, and the supposedly immutable glory of our planetary orbits, will manage to survive only as far into the future as it has already survived since the Sun formed. Not exactly a great comfort.

In light of these facts I think it's fair to say that the notion of the clockwork nature of the heavens now counts as one of the greatest illusions in the history of science, brought about by our limited perceptions, and by the way in which we happen to have developed our cosmic models. Indeed, even the very simplest of systems—a star and a single planet—is never truly unchanging. The star is not a singular point, as models based on Newton's laws typically assume. It is a large and stratified object that may not be perfectly spherical or even of a constant mass.

A star can shed some of its material over time as it bleeds photons and particles into space, and its gaseous outer shell can be yanked at and distorted by the tidal pull of a planet, even if only minutely. The planet itself is also not a compact point; it is close to—but unlikely to be—perfectly spherical. Like any large rocky or gaseous object, it will also be layered inside like a colossal onion, made of compounds of differing densities and viscosities.

A planet, as I've discussed, may leak substantial amounts of its atmosphere into space, and it too will feel the pull of tides from the star's gravitational field. The gentle friction as it is kneaded by these forces slowly releases energy that may never be retrieved as it radiates away to the cosmos. That energy is ultimately drained from the planet's spin and from its orbit. And even the orientation of its spin axis can change with time. Altogether, whether we like it or not, the "simple" star-and-planet system will evolve.

Another basic example of a two-body arrangement is our own Earth-Moon system. Even if we magically isolate these bodies from the influence of the Sun's gravity, we find that nothing is truly fixed. When the Moon formed, from what we think was a great collision in the rough-and-tumble of our embryonic solar system, it found itself in orbit around a fast-spinning Earth. Today the Earth's twenty-four-hour rate of spin still readily exceeds the twenty-seven-day orbital period of the Moon, but it will not do so forever.

The tides that the Moon's gravity raises in our oceans and landmasses are manifest as great shallow bulges of matter. But during the time these bulges lift up toward the Moon, our restless planet spins ever onward, carrying them ahead of the Moon's position above us. The result is an uneven gravitational pull on the Moon. The racing bulge tugs the Moon not toward the Earth, but rather along in its path. As a result, the Moon gets swung to a higher orbit, but its pull simultaneously slows Earth's spin. On paltry human timescales these are tiny effects, but they are measurable, and we've managed to perform just such an experiment.

When the Apollo astronauts visited the lunar surface in the late 1960s and early 1970s, some of the items they left behind were specially designed mirrors. Inclined to face the Earth, these mirrors, and their counterparts left by Soviet lunar missions, are used to bounce laser beams off the Moon and determine its range to very high precision. It's a decidedly tricky measurement. Owing to the distance and the spread and scatter of light through our atmosphere and off the Moon-based reflectors, only about one in every hundred thousand trillion photons of light actually makes it back to be detected by us.

Nonetheless, the precise color and timing of the laser pulses allows our electronic instruments to pick up this feeble return signal and to time its arrival. We also know precisely how fast light travels, and we know how to take care of the extra effects of lunar orbital wobbles and Einstein's relativity. As a result, we can convert this total round-trip time for the flight of the photons (about two and a half seconds) into a measure of distance. What we find is that every year the Moon recedes from us by almost 4 centimeters, or by 0.0000000008 percent of its present distance, and the Earth's day slows by 0.0000015 seconds.

These are minute quantities, but the system is clearly not immutable. There is nothing permanent about its orbital dance steps. Indeed, paleontological records of ancient shorelines and tidally deposited minerals and fossils provide us with evidence that our planetary spin ran differently in the past. Some 600 million years ago it appears that the Earth's day length was only about twenty-one hours; our rotation has slowed by a total of three hours since waves washed ashore on those distant beaches.

So in many respects the perfection of Newton's equations describing the motion of planets is a consequence of making some significant approximations. Even Einstein's beautiful generalizations of these equations don't capture the messy details. While mathematics still rules the universe, making predictions is seldom straightforward because of the accumulation of effects that we might at first overlook—n-body effects that can sometimes catapult planets to disaster or rearrange an entire system.

All of these discoveries bring us back to a central question in the quest to understand our cosmic significance, because orbital properties represent another marker by which to compare our solar system with others. Indeed, the fact that the stability of planetary pathways is illusory opens up a new vantage point, just like when Kepler realized that planetary orbits were elliptical, and unleashed an enormous array of possibilities for their configurations.

It means that there is another vital characteristic of any planetary system, an additional feature to know about. Beyond the instantaneous configuration of orbits that we see is the question of what those orbits are going to do in the future, or what they have done in the past. In other words, a planetary system cannot be fully known by a single snapshot in time. It is an evolving, varying, and potentially chaotic beast.

If Copernicus had been presented with these facts, he might have given up trying to configure the heavens on the spot. After all, if the colossal revolution of displacing the Earth from the center of the cosmos wasn't sufficient to describe the totality of celestial reality, then how could we ever hope to understand the nature of things? Luckily for us, though, this extra characteristic is also an opportunity, because it might just provide us with another vitally important way to categorize our solar system.

In the previous chapter I introduced you to the league of extraordinary planets, and their remarkable abundance and range of properties, including the seemingly endless numbers of combinations and permutations for their orbital pathways. I also hinted at the reason for some of these configurations: pasts full of change and variation. Now we've come almost full circle. By discovering that our own solar system lives on the border of chaos, we have equipped ourselves to return to these other worlds, the exoplanets, and ask how they came to be the way they are.

The answer is going to reveal yet another clue about our status among all of this planetary chaos.

To explore the ballet of exoplanets, we need to dip our toes back into the science of simulation, the computer-driven models of gravitational forces between bodies. I'll confess that I'm the sort of person who gets excited by gizmos and gadgets—especially if they seem to offer the perfect solution to a vexing problem. When facing any household crisis, nothing beats the sense of satisfaction of knowing which tool to pull from your toolbox, where it was carefully placed in anticipation of just such a need. Such moments are cause for a celebratory cup of tea and philosophical munching of a cookie, often while something else out of sight and mind goes horribly wrong.

Some tools in science are equally satisfying, even if they are not a panacea for everything. Computer systems and programs that mimic the gravitational dynamics of objects are, I think, high-ranking in this toolset. The history of the development of these remarkable simulators and calculus-crunching machines is fascinating, but I'll leave that story for another day, because I want to focus on how they're leading us to a radically new vision of the nature of all planetary systems, not just our own.

The first time I played around with one of these exquisitely crafted computer codes, made freely available by a talented dynamicist, I could barely wait until the next morning when I could check on its progress. I was eager to see where my imagined worlds had taken themselves, what orbital mischief they had been at during their virtual electronic cycles.

It was such guilty fun to plot out each planet's history, millions of years of gravitationally driven movement codified into simple patterns and paths on my screen. Perhaps there is also a taste of megalomania in this opportunity to wield a godlike power over whole solar systems, the life and death of worlds—all created by your own hands!—played out in a speck on a microscope slide.

Whatever the reason, the allure is strong, and the scientific culture surrounding those who devote themselves to the challenge of taming the endless variations of gravitational interactions is distinctive and flavorful.* By simulating a seemingly endless array of real and imaginary planetary systems, scientists can examine hypotheses that would otherwise be nearly impossible to study. Critically, over the past decade a number of researchers have used these simulators to investigate how hypothetical freshly formed planetary systems behave.

As I've talked about earlier, we think that the basic mechanism of planet formation is through the coalescence or coagulation of material out of great disks of gas and dust that surround baby stars. But these disks are relatively short-lived, almost like the last few swirls of bubbles as you let the steaming soapy water out of the bathtub—except it's not the plunge down the drain but the intense energy of starlight that helps finish them off. While planets are assembling within these disks, they are more or less stuck, held in orbital paths by the mass of gas and dust around them; but when that stuff boils away, the planets are left feeling only the pull of each other, and can begin to determine their own future pathways.

---

* Even the language of orbital dynamics is very special. People talk about such things as resonances, precessions, librations, osculating elements, apsidal alignments, arguments of pericenter, harmonics, secular perturbations, and always, always, there is mention of chaos. Many of these phrases originate from other eras, reaching back through the 1800s and 1700s to Newton, Laplace, Lagrange, and other mathematical minds. It's a heavy armory of rich mathematical concepts, and its application to the emerging discoveries of exoplanetary science is revealing many surprises.

What many scientists have realized is that in this situation planetary systems might go through a period of youthful chaos, or instability, strong enough to result in the wholesale rearrangement of orbits and even the destruction or ejection of entire worlds. It's like a prehistoric, extreme version of the chaos that our own solar system may gradually approach in the future.

This might seem like just an untestable fantasy, but as more and more computer simulations have been undertaken to study the enormous range of possible outcomes of planetary instability, a striking pattern has emerged. Unstable young planetary systems eventually become the same types of exoplanetary systems that we find in the real universe, complete with strongly elliptical orbits and hot Jupiters. They are also responsible for ejecting worlds to interstellar space, where we have indeed picked up their telltale signatures.

The computer simulation of this process is almost magical. Take a thousand make-believe systems, throw them in the electronic conjuror's hat, let their orbital trajectories evolve untouched for the equivalent of a million to a hundred million years, and then see what planetary configurations are left. This remnant is a very good match, statistically speaking, for the properties of the hundreds and thousands of exoplanetary systems that we've discovered.

Here's another way to conceptualize all of this. A young and unstable planetary system can be thought of as being "hot," just as a cup of coffee or tea is hot. And like those beverages, anything that is hot will tend to cool off. In a cup of liquid, the cooling takes place as the hottest, fastest-moving molecules evaporate off and as thermal energy radiates away as infrared light. In an unstable planetary system the "cooling" occurs as some planets are ejected to interstellar space, or plunged into the central star, or driven into collision with each other. So a "hot" system with many planets becomes a "cool" system with few planets—the crowded instability of youth eventually relaxes into the roomy stability of middle age.

How often does this really happen in our galaxy? How many systems were dynamically hot in their youth? The present studies all indicate that around 75 percent—the great majority—of planetary systems go through an early episode of severe instability. Such a level of disarray is a remarkable proposition, but it seems to be a very good match to reality. Not only is the galaxy and the universe filled with planets around stars, but most of these planets exist in systems that are now configured quite differently than they were at birth.

It makes me think of the atomists of ancient Greece conjecturing a plurality of worlds. Except those old ideas now have to be modified to include a diversity of dynamical evolution, from hot to cold. Every planetary system comes with its unique story of lost or destroyed worlds, followed by periods of calm. However, nothing is truly guaranteed in the nonlinear, dog-prodding realm of orbital mechanics, and today's peace may give way to a chaotic future.

This is one of the most shocking developments in planetary science of the past two decades. Although it's not so surprising to find that a few systems have undergone episodes of "hot" orbital instability, to find out that this has affected *more than two-thirds* of them forces a genuine paradigm shift in the way we characterize planets. In part, this behavior is a direct consequence of the abundance of planets we see everywhere, a plenitude that implies that planets form efficiently. The more young worlds are packed around a new star, the more likely it is that they'll find themselves heading for chaos as their gravitational pull tugs at their siblings.

This paradigm also returns the focus to our own circumstances. We've discovered that the solar system is touched by a dab of chaos. But compared with so many other systems, it appears to have been relatively cool in dynamical terms. The orbits of all major planets are today only mildly elliptical, and the arrangement of worlds is rather sedate: smaller rocky planets in the inner system, and the giants in the outer system.

This is not to say that our youthful solar system didn't experience some rejiggering. A leading theory developed by many scientists attempts to explain the present configuration of the giant planets, and the distribution of smaller bodies in the asteroid belt and the distant Kuiper Belt, by invoking large shifts in the sizes of the orbits of Uranus and Neptune. In fact, this theory proposes that Uranus and Neptune once swapped places as both worlds migrated outward from an originally much more compact configuration. As this rearrangement played out, Uranus ended up in its present orbit and Neptune crossed its path, moving to become the farthest major planet from the Sun.

During this episode Saturn's orbit would shift a little outward to its current place, while massive Jupiter shifted a little inward. As in any mechanical system, you can't move things without an exchange of force, some kind of leverage. In this case part of the leverage, or exchange, would be the redistribution of much smaller objects—tens of thousands of icy chunks and rocky asteroids that could each impart a tiny push or pull as they interacted gravitationally with the larger worlds.

This orbital settling would have taken place some 4 billion years ago, just a few hundred million years after the dispersal of our disk of protoplanetary gas and dust. The last nudges of rearrangement would have helped perform a final housecleaning of the smaller pieces of material left over from the formation of the main planets. But if it did take place, it would rank quite low on the scale of dynamical instability, placing our solar system in the lukewarm rather than scorching hot category.

Another hypothesis for these early stages in the solar system's history was put forward recently by the dynamicist David Nesvorný, and in some respects it argues for our system to be more active, less unusual, less significant. In this picture the young solar system contained not four giant planets, but five. The fifth planet could have been an icy giant, perhaps intermediate between Neptune and Uranus in mass, and in an orbit somewhere beyond that of Saturn.

The formation of an object like this out of the gunk of gas and dust around our baby Sun is certainly plausible, and could add a little more spice to the solar system's orbital history. Nesvorný's simulations of the subsequent evolution of such a system typically result in the fifth giant being given the gravitational heave-ho by Jupiter, which ejects it all the way out to interstellar space. But the arrangement of our major planets that's left behind in such simulations is often a good statistical match to the configuration that we see now. In other words (and perhaps counterintuitively), the presence of this extra planet could be just what the doctor ordered. Having had a fifth giant planet, now lost, seems to increase the likelihood of our youthful solar system ending up looking like the present solar system.

This is certainly an interesting twist, and a great reminder that we still don't know exactly what happened 4 billion years ago in our own system. Perhaps the current, rather subdued dynamical state of our planets owes something to a rather more violent, hotter, dynamical past. Perhaps we flung a sister world out into the void. The brutal indifference of natural selection may work on planets, too.

Whatever happened in the solar system's past was still probably relatively mild compared with the events in a majority of planetary systems, and the comparatively circular and well-behaved arrangement of our planets today is a reflection of that. All of which really brings us to the crux of my discussion, and it's straightforward: the architecture of our solar system provides us with a special fingerprint that, for the first time, allows us to make a moderately robust statement about its degree of uniqueness.

The simplest pieces of this fingerprint are the shapes and orientations of the planetary orbits, and the placement and variety of planets. From the configuration of our orbits alone, it seems like a good bet that our solar system belongs to the 25 percent or so of planetary systems that have never been particularly chaotic in the past. Our system also contains no worlds in the range between the mass of the Earth and the masses of the ice giants, Uranus and Neptune. These giant planets are about eighty and one hundred times the mass of the Earth, respectively. That means there is quite a gap in mass between our small rocky worlds and anything bigger.

We now think that planets in this middle range, from super-Earths to baby Neptunes, are among the most numerous types of planets of all—perhaps outnumbering giant worlds by a factor of four or more. Yet there are no examples around the Sun, and we might have never conceived of such worlds if we hadn't found them around other stars. Current estimates are that more than 60 percent of other suns harbor at least one of these middle-range planets.

It's admittedly hard to combine all of these statistics in a way that's rigorous. For example, we don't really know whether the dynamical instability of systems is also connected in some way to their propensity for forming super-Earths and mini-Neptunes. So it may be like seeing lots of flowers in one corner of a garden. Perhaps they're there solely by luck, or perhaps they're abundant because that corner is well tended by an unseen gardener. Nonetheless, it is quite clear that in these terms the solar system is still somewhat unusual, somewhat of an outlier, a member of a minority.

Let's assume for the sake of simplicity that the form of the orbital architecture and the types of planets in a system are not strongly related. This is a supposition that is likely incorrect at some level, but making this approximation sidesteps the need for a more sophisticated analysis that probably wouldn't alter the broader conclusions. So, we can combine the odds to conclude that we live in a solar system that belongs to a 10 percent club, at most. To take this further we might add some other simple facts to this statistical recipe.

For example, I've talked about how most stars in our galaxy are smaller than the Sun—some 75 percent are less massive. These stars are also host to innumerable planets that appear to follow the common dynamical rules of hot youth and cool middle age. So if we were to tentatively merge our statistics a little further, we might be tempted to say that our solar system is more likely to belong to a 2 or 3 percent club—a certain type of star with a certain mix and arrangement of planets. This is not mathematically rigorous, but it is based on real numbers, and represents a key part of our quest to understand our cosmic significance. Taken as a whole, our solar system is unusual.

I've also talked about the question of a planet having a temperate surface environment, capable of harboring liquid water. Astronomers love to take this idea and talk about "habitable zones" around stars—orbital regions where a planet's temperature may fit neatly between the freezing and boiling points of water. At face value, this also significantly reduces the size of the club that our solar system and Earth belong to—demanding that worlds orbit at a just-right distance from their parent stars.

It's very difficult to accurately assess this population, and I'm loath to do so. In truth it hinges on a multitude of factors, such as planetary compositions, atmospheres, and environmental stability, as I discussed in the last chapter. We're also still struggling to understand the basics of our own planetary climate. We think that the Sun was 30 percent fainter 4 billion years ago, yet geological evidence points to liquid water on Earth's surface back then. The problem is that we're not entirely sure how this could be. Even a potent mix of greenhouse gases in the young Earth's atmosphere would be hard-pressed to warm the surface enough and not reveal itself through the rock record. Some researchers have even suggested that the fundamental shape, size, and optical properties of clouds (yes, clouds) were different billions of years ago. Different clouds could have made Earth a less reflective planet, capable of absorbing more of the Sun's warming energy.

We also have increasingly good evidence that Mars, a planet just outside the expected temperate orbital zone around the Sun, once harbored plenty of liquid water. It may not have done so for very long in geological terms, perhaps undergoing brief episodes of wetness, but nonetheless, at times conditions there have been far more amenable to life than they are now.

The upshot is that in terms of temperate environments, it's tricky to know how to evaluate the unusualness of our solar system. I'd say that at this time, given our level of knowledge, we can't put any reliable number on the fraction of systems with planets in temperate zones, because those temperate zones themselves are fickle things. However, adding in the solar system's history of temperate planetary environments to our calculation conceivably places it in a club of less than 1 percent of all possible planetary systems.

But that is all statistics. What characteristics actually determine the unique and detailed nature of an individual system? Why are the odds the way they are for systems to form dynamically hot or cold, and with or without certain planetary flavors? And what sets in motion the events that end up producing a solar system like ours and a planet specifically like Earth?

Some of the answers undoubtedly lie in the general physics of gravitational systems, and in the intrinsic attraction of gases and particles that swirl around a baby star as it gathers itself up out of the chill soup of interstellar material. But a big piece of the puzzle, a very big piece, seems to come from sheer, blind, unadulterated chance.

Astronomers talk about the formation of planets as a stochastic process, meaning that while there are underlying, predictable physical processes, the final outcome is inherently nondeterministic; there is a random element. I can tell you what's going on in general—matter is orbiting, colliding, coalescing, and objects are interacting, scattering, growing, and breaking—but I cannot predict what will happen with every new world or chunk of material. It's just like the impossible n-body problem.

One of the best examples of this stares us in the face almost every night. The Moon, our Luna, as I've mentioned before, is likely the result of a cosmic collision between an earlier version of the Earth and another embryonic planetary body. The theory that best fits our current understanding of the nature of the Moon and Earth is that approximately 4.5 billion years ago another world the size of Mars collided with the proto-Earth. This unfortunate body is known as Theia and may have formed in the same orbital zone as the proto-Earth, but separated in its position around the Sun.

Over time, the variations of gravitational pulls would have brought these two nascent objects together, until they eventually cracked into each other like a pair of colossal tumbling boulders. In the resulting mess, Luna would have quickly coalesced from the debris that encircled the Earth—a mix of what used to be Theia with the stripped and pummeled outer layers of what used to be proto-Earth.

An event like this may not be at all unusual as a planetary system forms. It represents the tail end of the violence and orbital jostling that we think plays a central role in putting the finishing touches to small rocky planets. But it is also not a foregone conclusion—it is part of a highly stochastic set of events, where the likelihood of a particular outcome is difficult to predict. The Earth and the Moon may be a relatively common type of planet-satellite configuration, but it is never going to be guaranteed in any specific case.

This characteristic is just another aspect of the nonlinear, chaotic nature of a planetary system. Except there is an added factor: fine details determining the outcome that have less to do with gravitational rules and more to do with the accidental sizes and compositions of planets. For example, the physical collision of two objects doesn't just depend on whether they pass close to each other; it depends on their girth: will they be wide enough to clip each other? And if so, will that collision result in their merger into something new, or just break them into smaller pieces?

So we face a huge challenge if we want to try to point our finger at the pathway of cause-and-effect that leads from cosmic gas and dust to a planet like the Earth, but that's the way things are. At the same time, though, it's also important to recognize that just because the route to a final state is unpredictable and random, that doesn't automatically mean that that final state is unlikely. I can't emphasize this paradox enough, because it's a characteristic we'll encounter in this tale when talking about more than just planetary systems.

One way to think about this aspect of the evolution of natural systems is to imagine that you're standing at the entrance to a thick forest that you need to get through. You may have numerous paths to choose from, and perhaps 90 percent of them will lead you to someplace on the other side of the trees, while only 10 percent will leave you going in perpetual circles. The odds are good that you'll make it through, but you still have no choice but to pick one path at random. And even if you're lucky, you'll emerge in a slightly different place each time. Making planets is a pretty similar process, and as we'll see, this could be true for the phenomenon of life as well.

Having journeyed unscathed through this synopsis of the dynamical nature of celestial mechanics, you may be tempted to breathe a sigh of relief. But I'm afraid there's another aspect of planetary systems that gives them one more layer of complexity. Whether planets, asteroids, comets, and dust orbit a single star or multiple stars, we've always tended to think of these systems as closed boxes. They have appeared to be isolated ecosystems—save for the occasional ejection of the odd planet or two along the way. It turns out this may not be correct.

The inner reaches of our solar system at first appear relatively tight-knit when it comes to invasion by pieces of solid matter. Perhaps a little interstellar dust makes its way in now and then, but otherwise the only significant interlopers to the neighborhood are certain types of comets. Back when I described the grand arrangement of our system, I talked about the Oort Cloud, a postulated outermost reservoir of hundreds of billions of icy objects flung into distant and slow orbits during the heyday of our solar system's youth. Every so often one of these pieces of ancient material tumbles onto a trajectory that brings it inward, and these events produce a special population of so-called long-period comets. Seeing these comets provides a critical clue that the Oort Cloud exists, stretching out to almost a light-year from us, well on the way to the next star.

But there's long been a bit of a problem with this hypothesis. There are simply too many of these long-period comets to be accounted for by leftovers from the formation of the solar system. A homegrown Oort Cloud, consisting solely of material flung outward by our nascent planetary system, could not contain enough cometary bodies to account for what we see.

This inconsistency has been puzzling astronomers for quite a while, but recently the scientist Hal Levison and his colleagues came up with a plausible theory. It hinges on something else that we've already encountered—the birth of our sun and its planets in among a collection of sister suns, now long dispersed or lost within the galaxy.

Levison and his colleagues have applied gravitational computer simulations to the problem, tracing the orbital trajectories not only of planets around stars in a cluster of sister stars, but also the paths of their icy Oort-like detritus. What they find is pretty amazing. Since the birth grouping of stars is extremely compact, what takes place is something akin to a mad cartoon scuffle.

Many of the icy bits and pieces from around individual stars get stripped away by gravity, forming a great cloud of shared material surrounding the whole stellar family—the kind of exaggerated blur of stuff any TV aficionado would recognize, with the occasional limb or exclamation point emerging. Subsequently, as the stars continue to move around in this cloud they can greedily sweep up that material again. Sometimes they also pass exceptionally close to another star, in which case they can steal even vaster numbers of these tiny pieces, clutching onto them with their gravitational tendrils.

The upshot of all this brawling is that stars have the opportunity to accumulate far more material into their Oort clouds than they might in splendid isolation—enough to explain what we see in the solar system. We don't yet know for sure whether this is what happens, but it's an appealing solution to what is otherwise a mystery.

The piece of it that I want to emphasize, because it matters the most to us in our quest to understand our cosmic significance, is that if correct, this research suggests that as much as 90 *percent* of our Oort Cloud is extrasolar in origin. Our solar system's outer limbs are *not* its own; rather, they are borrowed and stolen matter from our wild youth. Equally, our home-grown icy detritus has largely gone elsewhere, stolen and borrowed by other stars, or just left for dead in deepest interstellar space. In short, the solar system is a leaky boat that has taken on a substantial amount of alien bilge.

What brings this literally very close to home is that long-period comets, those icy bodies from the Oort Cloud, can reach all the way to within the orbits of Jupiter and even the Earth. When they do, they behave as comets must: solar radiation causes their volatile ices to turn to gas, which boils into interplanetary space, carrying off the dust that is also part of their matrix. This has been going on for billions of years.

If Hal Levison and his colleagues are correct, our local environment is regularly being polluted by the chemistry of other solar systems. Not only is our solar system a fickle thing; its current physical contents may be far from native.

❧

Imagine for a moment that Aristotle, or Ptolemy, or Copernicus, or Kepler and Galileo had unearthed such facts about the world around them. So much would have been different! In particular, these newly understood characteristics of our solar system further demolish any lingering notion that we exist in a long-lived or perfectly tuned place. It may be quite low on the scale of chaotic realms, but it certainly isn't at the bottom, and it has been and still is in flux.

Seen through the lens of orbital dynamics, our place in the cosmos is *startlingly* different from what was posited by any of these earlier thinkers and scientists. Simply displacing the Earth from the center of the universe barely scratches the surface in terms of finding a way to adequately measure our cosmic significance. We are riding a single speck that's afloat on a wild ocean of pathways and possibilities. But at the same time, it is not just any old speck. We now know that our solar system is at least somewhat unusual, and we have the numbers to back that up.

Now, of course one could argue that it doesn't matter that our perch is impermanent or even unusual. Individual human lifetimes operate on a completely different clock than does the cosmos. Even the entire history of the evolutionary development of mammalian life over the past 200 million years is a brief event compared to the time line of stars and planetary systems.

But our species didn't spring from a barren world. The Earth (as we'll see) has an intertwined history with life that stretches back across nearly all of its 4.5 billion years. Without this backdrop, none of us would likely exist. Yet that local chemical and biological history is *also* nonlinear, veering into chaos at times, just like our planetary orbits, and ultimately governed by the same underlying mathematical properties that took a bite out of Henri Poincaré's bank balance.

Most of that complex biochemical story takes place in a different layer of our universe, a layer deep in the microcosm. This is where we have to go next, because in order to find the connection between an unusual solar system and the existence of life, we need to make sure we understand what life really is, as well as its relationship to planets, and to the cosmos beyond.

5

SUGAR AND SPICE

It's startling to realize that until very, very recently we probably knew more about the universe beyond Earth's atmosphere than we did about the enormously complex nature of terrestrial biology. But now, four centuries after the invention of telescopes and microscopes and the first sightings made by Antony van Leeuwenhoek, the veil is lifting. There is a whole other world right under our noses, a hidden dimension, wrapped up out of everyday sight—the intricate, swarming world of molecules, membranes, and cells that compose life. In this weird and wonderful place we'll find some of the greatest clues yet to the connections between life and the basic properties of the cosmos.

Our understanding of this terrestrial microcosm has a long way to go before it's complete, but we've already managed to discover many of its vital characteristics. The first concerns our macroscopic biological world as well. We now think that there are three major domains of Earth's living things, distinct blueprints for organisms: the bacteria, the archaea ("old ones"), and the eukaryotes. (It is still up for debate exactly where to place viruses, or whether they are even alive, so for now they're put somewhere off to the side.) These three forms of life are readily distinguished by the basic architecture of their cells, as well as by their genetic codes.

In a nutshell, bacteria and archaea are "simple," small, single-celled organisms. They can survive as individuals, but more often operate as colonies. Their genetic material is loosely contained, and their cells tend not to contain any additional complex internal structures, or what are called organelles. By contrast, cells of eukaryotes are much larger and more complex, contain organelles, and keep their genetic material protectively bound up in a nucleus. As we'll discuss in much more detail, an evolutionary ancestry of symbiosis (two or more different organisms working together) has apparently bequeathed the eukaryotes an array of added abilities, including potent energy-producing mechanisms and the great trick of multicellularity. Humans and all animals, plants, and insects, and even the humble fungi, are eukaryotes. However, we eukaryotes remain critically reliant on symbiotic partners from the simple-celled kingdoms of life, as we'll see when we explore the human microbiome.

Simple-celled life-forms (collectively, the prokaryotes) represent the most ancient forms of life on the planet. Bacteria are just a few micrometers wide. They come in an array of shapes, ranging from spheres to tubes, rods, and spirals, that sometimes propel themselves with whiplike spinning tails called flagella; they are highly diverse. The denizens of the other ancient domain, the equally minuscule archaea, have dealt us a lesson in abject humility. Until the late 1970s we didn't recognize them as a truly distinct form of life, assuming instead that they were just another type of wiggling bacteria. But they're not. Many of their cellular structures are fundamentally different, and even their propelling tails are built differently than those of bacteria. They also have a proclivity for "living off the land" in an incredibly wide range of environments. They can do this by consuming raw, simple chemical compounds—a feature that further supports the notion that their lineage goes a long, long way back, to a time when the only food was inorganic matter.

It's too easy to imagine that such life-forms are primitive in function as well as ancient in form. Far from it! Each tiny individual is a piece of intricate natural machinery. Even their apparently simple tails are powered by the sophisticated molecular equivalent of an electric motor spinning at hundreds of revolutions per minute. As we'll see, the full scope of their abilities goes much deeper still.

There are also a lot of them. Our current estimates are that the planet Earth harbors in excess of a million trillion trillion (10^{30}) single-celled organisms. Their genetic diversity is staggering, with at least ten million distinct species and likely many more. In the past thirty to forty years we've discovered that many of these microbes thrive in environments that we cannot tolerate—places of extreme temperature, pressure, chemical toxicity, and sometimes all three. This toughness lets microbial life occupy almost every nook and cranny, temperate or hostile, that the planet has to offer. These organisms are not only the most diverse and populous on Earth by far, but also a major fraction of the planetary biomass.

Most of this living sprawl doesn't even occupy the immediate surface of the planet. Marine environments, especially the upper layers of the ocean, are full of microbial organisms. Deeper down, the sediments and rock of ocean floors provide a virtual megacity for life that covers 70 percent of the planet. Much is sparse and slow-living. But close to volcanic mid-ocean ridges, which span an interconnected length of more than 37,000 miles around the planet, organisms can bloom into oasis-like communities. On the continental landmasses, organisms exist within the soil and ice, and within a microscopic jungle of fissures and cracks that extends into the planetary crust. Evidence of microbes has even been found within the glassy basalts of young volcanic mountain cones, where they live directly off the rock.

If someone just a hundred years ago was asked what constituted the bulk of life on Earth, his or her answer would probably involve plants or insects—certainly not bacteria, and certainly not the million trillion trillion cells of microbial life that we now know live mostly hidden out of sight in the subsurface zones of the planet. Yet this burgeoning and pervasive population is key to our existence and key to the question of our significance. It's these bacteria and archaea that harbor the secrets of life on Earth—the master plans for harvesting energy and materials, building biological structures, and employing some of the most amazing chemical trickery we know of. In fact, the most immediately noticeable aspects of our world—from

the atmosphere to oceans and the chemistry of the rocks and soil—
have been mindlessly but brilliantly engineered for the past 4 billion
years by these same microbial denizens.

❧

To grasp the sheer scale of life's integration into our planetary system takes some readjustment. For me, the moment when my previously narrow ideas about the nature of life on this planet took the
biggest hit came in 2008. It happened when I read an article in
Science written by the biologist and biological oceanographer Paul
Falkowski and the marine microbiologists Tom Fenchel and Edward Delong. The title of this piece was "The Microbial Engines
That Drive Earth's Biogeochemical Cycles," a straightforward-
sounding description that belies the profundity of what's discussed.

What are these "microbial engines"? Mechanically, most are in
the form of complicated bindings of molecules known as proteins.
High school biology teaches us that proteins, in turn, consist of chain-
like sequences and folded assemblages of simpler molecules called
amino acids. In terrestrial biochemistry, the relevant amino acids are
a selection of twenty molecular structures, each containing between
ten and twenty-seven atoms of elements such as carbon, hydrogen,
oxygen, nitrogen, and sulfur. They're basic building blocks—the Lego
pieces—used within cells, and the genetic code inside all living
things provides the instructions for putting these pieces together.

The proteins that life builds from amino acids are the workhorses
of biochemistry. They can act as catalysts to encourage chemical reactions, and they can come together to form larger structures. If they
combine into what are known as "multimeric protein complexes,"
they become full-fledged molecular machines—sophisticated pieces
of natural engineering produced by the restless actions of selection
and evolution. They are the working parts of all life. Indeed, in some
single-celled microbial life, proteins can account for as much as
50 percent of the dry mass of an organism.

The reason that some of these protein-based structures earn the
title of "engine" is that they are engaged in the root functions of

metabolism—the production of usable chemical energy and the synthesis of new compounds, the very processes that drive each and every organism.

Again, this takes us back to the rudiments of high school science: What's the fuel? What powers these engines? Ultimately it all boils down to the movement and transfer of two of nature's most fundamental physical units—electrons and protons. The chemistry of life is sustained by the exchange and flux of these electrically charged particles, in what are known as reduction and oxidation reactions.

These reactions can sometimes happen all by themselves if the right molecules get close enough to each other with the right energetic kick. For example, with some heat, methane can combust in oxygen. It's a reaction we've all witnessed in kitchens when cooking with gas or in school labs when firing up Bunsen burners. The end result is that atoms of carbon and hydrogen bond with oxygen, and lose electrons in the process. (In fact, the term "oxidation" is a little archaic—what's really going on in this kind of reaction is that atoms are losing or transferring electrons.) And the transfer of charged particles means that there is a flow of energy that can be tapped into to power other processes.

But not all reactions are as spontaneous—they need further encouragement. So that's what life does: its molecular engines piggyback onto reactions, catalyze them, and extract some of the electrical energy for life's own purposes, often storing that energy in other molecules that transport it into the rest of an organism's cell or cells. This is what powers life on Earth. And in fact the molecular engines don't just piggyback, they physically gather the chemical fuel and they engineer conditions to make the reactions take place—they *metabolize*.

There is an enormous catch, though. All such chemical reactions, such transfers of electrons or protons, convert a set of ingredients into a set of products. So if Earth had a finite supply of raw and reactive chemistry for life to use, it would, over time, run out. But the planet isn't static. Churning geophysical activity, from volcanoes to plate tectonics, recycles organic sediments and their chemical components back to the surface, and sunlight-driven chemical reactions in the atmosphere are continuously generating fresh ingredients.

The problem is that these processes are quite slow, taking millions of years to substantially replenish the chemical larder. Life originated at least 3.5 billion years ago and has survived since then, so it must have found another way to subsist while the Earth trundles slowly on—and it has. This was the moment of revelation I experienced while I was reading the study by Falkowski, Fenchel, and Delong. Their paper explains how the molecular engines of life have evolved to form a remarkable interlinked system—a system whereby microbial organisms catalyze multiple reduction or oxidation reactions in a number of *self-sustaining* cycles. In other words, molecular engines act as the firestarters for chains of repeating chemistry that might otherwise happen only slowly or not at all.

As a result of metabolism, elements such as hydrogen, carbon, nitrogen, oxygen, and sulfur are endlessly shuttled back and forth between molecules and places. Over time, the chemical structure of Earth's crusty surface and oceans is profoundly worked and reworked into something it could never have been without the existence of life. This is biogeochemistry. Nearly the entire environment that we experience on the Earth, from the oxygen we breathe to the composition of the ground beneath us, is the resultant equilibrium of all these interlinked and interlocked cycles. Of course, we're not separate from this system. Life like us belongs to the eukaryotic domain, with large, complex cells that are apparently the result of various instances of endosymbiosis—the assimilation of machinery from earlier, purely symbiotic relationships among single-celled organisms. This complex-celled life relies almost exclusively on oxygen respiration, and on a variety of molecular carbon energy sources. And that makes our oxygen-hungry bodies a significant component within the system of planetwide metabolic cycles.

❧

These interleaved and self-sustaining cycles are key to our quest not only to understand the connection of all life to the chemical and physical fabric of the universe, but also to try to place ourselves in this greater context. There is a finite number of these metabolic processes—at least of those that occur on Earth today. In principle, other types of chemical reactions could take place, but billions of years of evolution on Earth have converged onto this particular set.

It can help to think about these metabolic recipes as a variety of combinations of molecular "fuel" with molecular oxidants that "burn" this fuel. The most familiar metabolic pathways use such processes as oxygen respiration, fermentation, nitrogen fixation, oxygen-producing photosynthesis, and non-oxygen-producing photosynthesis. Others seem more exotic: types of sulfate respiration, nitrogen dioxide respiration, and even iron- or manganese-based respiration. There are bacteria and archaea that specialize in any one, or sometimes several, of the metabolic choices available. For example, the molecular engines inside certain types of archaea can combine carbon dioxide (an oxidant) with molecular hydrogen (a fuel) to produce methane and water. They can also pull apart molecules of acetic acid to make methane and carbon dioxide. Most of the methane available to us humans, and (let's be blunt) produced by us humans and many other animals, comes from busy little archaea. This type of metabolic process is known as methanogenesis.

Above all, reactions involving carbon fixation—converting simple inorganic carbon sources, such as carbon dioxide, into organic compounds—are centrally important to the global biosphere, for carbon chemistry is the very basis of life on Earth. All in all, we've discovered that there are ten central chemical processes that we think represent the metabolic profile of life on Earth. These are the sum of the ways in which electrical energy and raw materials are obtained for all organisms.

But the way all of these processes operate within a nested system of cycles that is shared across all species and spread around the entire planet—that's the wonderful thing. For example, the molecular engines that some archaea use to produce methane can be switched into reverse in other archaea and bacteria. They extract energy by pulling methane apart and turning it back into carbon dioxide and hydrogen. Someone's waste is another's food.

Most of the other processes are similarly reversible. If there isn't a single microbial species with the machinery to directly undo what another has done, there will be a diverse set of interactions that may span many different species to exploit the reverse process step by step. The participating organisms need not even live in proximity to each other, either in space or in time. Methane produced somewhere on the planet by one set of organisms, for example, may find entirely different consumers elsewhere and at different times of year.

It sounds suspiciously like a perpetual motion machine, in which one organism produces food for another that in return converts it back, all the while extracting energy. It would be, except that all of this planetwide metabolism is not a closed system. It's ultimately driven by two energy sources that I've already mentioned. First, the Earth is still hot inside from its violent formation and radioactive additives, and somewhere between 30 and 45 trillion watts of averaged geothermal and geochemical power is delivered up to its surface. Second, its surface also absorbs energy at a rate of about 90,000 trillion watts from the Sun. These energy inputs more than make up for any losses due to inefficiencies in the interwoven metabolic cycles of life.

It's a beautiful system, but understanding it is simply preparation for asking how all of this microbial engineering evolved, and in particular how it's survived the vicissitudes of a planetary environment for the past 3 to 4 billion years. Part of the answer must rest on exactly how a relatively small set of molecular engines, mostly protein complexes, have become encoded in the genetic material of single-celled microbial organisms.

From geological studies of rock chemistry, as well as from genetic studies, we are confident that most of the DNA codes for these machines are ancient. Some have left their fingerprints quite literally in stone, as mineralized rock strata produced by entire ecosystems that once shifted the chemical balance of Earth's oceans and atmospheres. All have also left a trail in the genetic sequences of modern life.

Several of the engines require significant genetic information to code their structures. Oxygen-producing photosynthesis, for example, is the most complicated natural energy transduction process known, involving multiple molecular assemblages described by more than 100 genes. Yet we have evidence that photosynthesis was a metabolic tool at least 3 billion years ago. Such sophisticated molecular workings clearly evolved early in life's history on the Earth.

Understanding the origins of all these metabolic processes is key to unraveling the origins of life itself, which remain somewhat of a mystery. That's not to say that theories and hypotheses don't abound. For example, some scientists claim that there are striking similarities between chemical and electrical gradients in cell walls and those possible in the chemical imbalances and microscopic mineral structures found in deep-sea hydrothermal vent systems. This could indicate a potential inorganic template for life's origins—in other words, a blueprint built by nothing more than geophysics and geochemistry.

Suggestive connections like these between early life and nonbiological mineral structures and chemistry are intriguing, but so far, we have no definitive proof of a link. Other ideas involve multiple stages of early organic chemistry, networks of reactions of amino acids brought on by catalysts like boron and molybdenum in a watery environment. The basic pieces of biology, from fatty lipids to the first ribosomal structures (helping perform the synthesis of proteins), might start with such reaction chains.

Indeed, pieces of terrestrial biology could have emerged from a variety of sources. In that case, we need to understand how biologically useful molecular parts can come together out of disparate origins and build something more resilient. Luckily, nature itself may be providing us with clues.

Microbes (and we presume their ancestors, too) are notorious for what's called horizontal gene transfer—the exchange of pieces of genetic material between species. (It's a bit like the exchange of business cards, or prototypes for some ingenious device.) As a result, the detective work of tracking down how, when, and where particular genes arose can be immensely compromised. But this promiscuous behavior had one vitally important outcome that must have helped direct the history of life. Widespread genetic sharing has apparently ensured that, in terms of critical genes, everything is pretty much everywhere.

If you go out on a boat to the deep ocean, draw up a sample of cold seawater, and bring it back to your lab, you'll typically discover types of bacteria or archaea in it that most certainly don't thrive at the marine surface. For example, you can find so-called thermophiles, organisms that require very high temperatures to metabolize and reproduce. Or perhaps you'll spot some other out-of-place organism. It doesn't matter how unlikely this cold ocean environment is for these life-forms—they'll still be in your sample.

Almost anywhere you go on Earth, you'll find the equivalent of a microbial genetic diaspora. Representatives of the majority of phyla are usually present, even if they don't like the conditions. There are exceptions; recent studies have suggested that Earth's polar regions may harbor certain bacteria that don't exist in any number elsewhere. But these are limitations that still leave microbial populations spread over very large geographical ranges.

It makes sense. Tiny organisms are readily dispersed and transported around the globe by air and water, and they've had a very long time to insinuate themselves into just about every corner. What is important, though, is that it's not just the microbes that spread the world over—it's also the set of genes that are the instructions for the molecular machinery of metabolism. This key group of genetic codes describes the engines that have, in effect, built the world as we know it. Falkowski and his colleagues dub it the "core planetary gene set," and that's an excellent name.

The fact that microbes carrying the core planetary gene set exist everywhere provides a ready explanation for how the fundamental metabolic processes of life have survived intact for billions of years. They simply keep spare copies of their code all over the place. For example, let's suppose a stray ten-kilometer-diameter asteroid smacks into the Earth with the power of about 100 trillion tons of TNT. That's your common garden-variety "dinosaur killer," like the one that impacted the Yucatán Peninsula and may have helped speed a mass extinction 65 million years ago. Or let's imagine that it's 570 million or more years ago, and much of the Earth's surface is frozen over during one of the periods dubbed a "Snowball Earth" episode. Untold numbers of organisms will perish and entire species will be wiped out, never to arise again in the history of the planet.

Yet somewhere on Earth there will always be bacteria or archaea carrying part of the core planetary gene set, the instructions for the molecular machinery of metabolism. Their microscopic bodies have seeped and wiggled into every available crack, hole, and ocean bottom, and even into the water droplets of clouds in the sky. As individuals they may not live long, but that doesn't matter—in their billions of trillions they serve as guardians of the genes across the ages. Indeed, some species carry more than just one of the core genes, regardless of which ones they use for their own metabolism.

It's not very poetic, but the situation is a beautiful analogue to a distributed computing network. Today when you download an electronic book or music file, or even take a photograph with your cell phone, you often keep in your hands only a copy. Another copy either already exists on your computer or is uploaded via the Internet to another storage device. And it doesn't stop there. These copies "in the cloud" are themselves copied onto different devices, often into mammoth server farms that may sit on opposite sides of a continent. In this way, barring the end of the world as we know it, the data remains safe. It doesn't matter if a few copies are damaged or destroyed by power outages or deranged hackers, because somewhere there is a duplicate.

Just as our computer systems are mindless carriers of the information we place on them, one could argue that microbes are simply carriers of the information for metabolic machinery across time and across Earth. We don't know exactly how resilient this storage is. It's easy to imagine it has flaws, as clearly over 3 or 4 billion years there may have been some hiccups. But overall it appears to have kept the plans for the central mechanisms of life secure.

It's also worth noting that the core planetary genes are themselves not necessarily perfect. The metabolic machinery that is produced from their code is often not as efficient or as simple as we might expect it to be from theoretical chemical models. For example, neither the molecular structures that carry out oxygen-producing photosynthesis nor the ones that handle nitrogen fixation are without handicaps. Natural photosynthesis is not as efficient as is *theoretically* possible, and today's nitrogen-fixing organisms can sometimes have to mitigate the hazards of reactive oxygen by making an excess of the fixing protein engines to ensure that enough are operational at any time. Yet the code for these mechanisms has remained essentially unchanged for billions of years. It seems that if something is good enough to do the job, that's all that matters.

So whatever the chemical origins of life were, once early life locked onto a good thing—a successful strategy—it largely held fast. That gives cause for optimism that the clues to that deep past have not been erased. It also, I think, provides us with a reasonable hypothesis to take forward. While specific pieces of life's metabolic machinery could clearly vary from place to place, world to world, the overall architecture of Earth's microbial system may be pointing to a universal pattern. In other words, the success of our core planetary gene set and its fabulous fail-safe system could represent the way any biosphere anywhere needs to operate in order to survive for a very long time. Any exoplanet with life may need to have the equivalent of its own core gene set and its own distributed backup system.

That brings us to the next piece in this story—the piece that links Earth's life to the cosmic order of things.

❧

The great swarm of molecular machinery on our planet all works with the same chemical building blocks, those Lego pieces. There are small variations, for sure: archaea make some use of certain "mirror" molecules—right-handed versions of amino acids that are otherwise exclusively left-handed in all other known life. But this is a structural variation rather than any difference in fundamental chemical composition. Any claims of life operating via some kind of truly alternative biochemistry are at present unsubstantiated, as I'll talk a little about in the next chapter.

Taking a cosmic view, none of this is very surprising, for the simple reason that the chemistry of life on Earth appears to be the same as the dominant chemistry of the universe. To explain that statement, let me take you on a quick trip to meet our ultimate ancestors: the molecules that made the universe, going back to a time just after the Big Bang 13.8 billion years ago.

In this cosmic dawn there was hydrogen. There was also helium, but in the cooling void of our young universe a few hundred thousand years after the Big Bang, reactive hydrogen was the element with the brightest future. Unlike the inert gas helium, which mostly stays in its unattached single-atom state, hydrogen had the greatest potential for forming molecules, with itself first of all—H_2—and molecular hydrogen is the key to making stars and heavy elements and, indeed, starting up all of chemistry. It's a little-mentioned secret that stellar astrophysics actually began with molecular chemistry.

The reason for this is that single hydrogen atoms whizzing around in the cosmos have few options for losing some of their energy of motion. If matter can't cool down, it can't form condensed structures like dust or stars. Even if hydrogen atoms bump into one another, it's very inefficient for them to cool down—that can happen only if they convert energy into photons that are emitted into space, and these simple atoms don't do that very readily. A hydrogen *molecule*, consisting of two protons bound by the mutual electrical attraction of two electrons, is a different story.

Molecular hydrogen is like a pair of balls joined by a spring: it can literally vibrate and rotate, which opens a new channel for losing thermal energy. Colliding molecules convert some of the energy of their motion into vibrations and rotations, and these in turn can dissipate the energy by releasing photons. These slightly squishy molecular springs can settle themselves faster than atoms that behave like hard billiard balls, and so they cool more quickly.

So once the universe started combining hydrogen into these simple molecules, gas could drop in temperature more rapidly. Cold gas doesn't resist the compressive effects of gravity as much, and so molecular hydrogen really did lead directly to the formation of the first generations of stars. As a result, it also set in motion the production of all heavier elements.

This isn't the only flavor of molecular hydrogen that the universe makes. If we survey space for molecular species we find that after the simple two-atom version of molecular hydrogen, the next most abundant molecule is a three-atom version with the grand name of "protonated molecular hydrogen," or H_3^+. It's simply three protons bound together by two electrons, and since it's missing a third electron it carries an overall positive electrical charge.

H_3^+ is remarkable. Like basic molecular hydrogen, it plays a critical role in cooling off gas. It's also highly reactive, and it is at the root of most so-called molecular-ion chemistry taking place in interstellar space. We even find its spectroscopic signature in places like the atmosphere of Jupiter. In many respects, if ordinary molecular hydrogen is the cosmic grandmother molecule, H_3^+ is the cosmic mother molecule.

If you map out the chemical reactions that H_3^+ helps initiate, they're very diverse. The production of water is one possible outcome. Another is hydrogen cyanide, a molecule that we may have reason to be wary of, but which can serve as a key ingredient for making amino acids, among other precursor biomolecules. Methanol, ethanol, and acetylene also come out of the sequences of reactions that H_3^+ initiates. And following the expanding pathways of possibilities, we find that they lead directly to the formation of longer

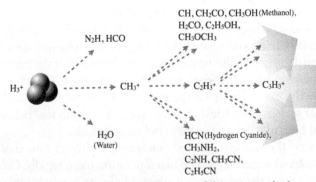

Figure 12: A schematic illustration of some of the compounds that are formed because of H_3^+. Reactions are possible that lead to progressively larger chains of carbon atoms and a diversity of molecules (*right*).

and longer chains of carbon-based molecules—the kinds of structures that are tantalizingly close to the molecules of biology.

This molecular triggering is a big clue to the ultimate basis of cosmic chemistry. Carbon, as I've mentioned, is an atom that happens to have a combination of outer electrons and overall size that makes it amenable to forming an astonishing array of molecular structures. With the help of H_3^+ it seems that just about anything is possible, within the thermodynamic constraints of the sparse chill of space.

Indeed, astronomers and astrochemists have found that the universe is *full* of carbon chemistry. More than 180 types of molecules have been directly identified in interstellar space by a variety of astronomical techniques, and more than 70 percent of these are carbon-based. The expectation is that this list is merely the tip of the iceberg, since larger and more varied molecules are likely there—just harder to identify, as their spectral fingerprints get increasingly messy.

An even richer chemistry appears in the denser tumult around forming stars and planetary systems. Immense amounts of molecular water are seen in many such places, together with a growing inventory of organic, carbon-based chemistry. We see alcohol molecules, sugar molecules, and evidence of precursors for amino acids such as glycine. All of this makes a great deal of sense in terms of what we know about chemistry. As we simultaneously develop mathematical models of what should be happening chemically in these environments, we find precisely the same reactions and compounds. Fundamental chemical theory predicts all that we see, and much more.

Put simply, we live in a universe of carbon chemistry that has deep roots in the most fundamental atomic physics and the original components of matter coming from the Big Bang. It's not at all hard to see how we might connect the dots between this knowledge, the rich but ancient brew we find in meteorites and comets, and the chemistry of life on Earth. With all these discoveries in hand, one would have to resort to contrived scenarios to propose any significant alternative pathway.

A real skeptic might say that all of this is entirely circumstantial, because we don't know what the steps are between simple carbon molecules and life. Yet this molecular connection offers a straightforward explanation for what's happened on Earth, an explanation that's perfectly aligned with our observations of the universe around us. Regardless of the details of how life began on the planet, the ubiquity of carbon chemistry in the cosmos removes any great surprise about terrestrial biochemistry. It's just part of the most varied and pervasive chemical network in all of existence.

Furthermore, the paleontological record on Earth points to microbial life having emerged very rapidly in geological terms. It appeared right on the tail of the last major acts of planetary assembly. We see that the chemical building blocks for life (sugars, alcohols, amino acids, and even more complex carbon structures) are already present in protoplanetary systems. This material can rain down on young planetary surfaces that are themselves great incubators for organic chemistry. In other words, the starter mix for life was easy to come by. This fact does not explain all subsequent steps, but it offers a clear signpost to the route taken.

We'll return to all these points as we evaluate questions of our quantifiable cosmic significance, but for me the biggest headlines are two. First, the geochemical state of the Earth has been reworked by a pervasive network of intertwined processes driven by trillions upon trillions of molecular engines that are the basis of microbial life, which in turn maintains the engine "blueprints" across time. Second, this entire microcosm appears to be directly related to the common carbon chemistry of the universe, and to the root origins of all physical and chemical structures in the hydrogen gas of the early universe.

I think there could still be open questions about the details of life's metabolic machinery here on Earth, and its degree of customization. This system has evolved both because of what's available on this planet and because of the vicissitudes of environment that help sculpt the processes of natural selection. In that sense there is a strong random element at play. One thing we do know is that the underlying chemical environment of the planet is ultimately determined by the earliest history of its formation from a condensing nebula, by the size of our parent star, and by the rough-and-tumble of planetary assembly. From what we've already seen of exoplanets, we expect that other worlds similar in size to the Earth may have enormous chemical and geophysical diversity.

So it's reasonable to suppose that not all of the metabolic processes that are used here on Earth would be plausible on other worlds. Equally, there could be reactions that work in those places but not here. A good example is one drawn from our studies of the Saturnian moon Titan. With temperatures of −290 degrees Fahrenheit, and a surface awash in liquid hydrocarbons, Titan has a chemistry substantially different from anyplace on Earth. However, there's at least one rather obvious metabolic process that could occur on Titan and that could provide life with useful energy. That's the reaction of hydrogen with acetylene. At our living temperatures on Earth this is a fairly explosive reaction, producing methane and a lot of noise. On cold Titan it would have to be catalyzed, but it would gently provide plentiful energy in return. Scientists have speculated about detecting life there by looking for signs of this metabolism in action. It may be a crazy-sounding idea, but it's within the bounds of possibility.

Despite the potential variations in specific metabolic details, there's little evidence to suggest that Earth's integrated system of metabolism and geochemical change is a fluke. On the contrary, as I've argued, looking beyond the fine print it seems to be a robust and plausible model for how *any* successful biosphere should operate.

So how do humans fit in? Life as we embody it has evolved out of the microcosm but is still completely integrated with it, and utterly reliant on it both for the global environment and for our individual functioning. This is true to a degree that we are only now beginning to fully understand, as I'll talk about next.

One of the most personally disturbing and revolutionary discoveries in modern biology is that we are not individuals in the way we thought we were. *We* are really "*us*"—collections of 10 trillion or so eukaryotic human cells that effectively serve as a scaffolding for a collection of *100* trillion individual microbes. The implications are staggering, and are rapidly transforming our view of human physiology and medicine. Welcome to the hidden world of the microbiome.

Most of us are not aware of this microscopic baggage through any direct experience. We don't noticeably slough off great wads of microbial life. But part of the reason that a simple inspection won't easily account for these organisms is that they are tiny, like the microbes that have found their way into every corner of our planet. A bacterial cell can be ten times smaller across than one of our cells, and the accumulated weight of our microbial passengers is thought to make up a total of only one or two pounds in an adult human.

That's barely 1 percent of our personal biomass. Yet, once again, this is the microcosm, the unsuspected world that Antony van Leeuwenhoek first glimpsed in 1674 through his ingenious microscopes; the world that it has taken science another four centuries to truly appreciate. Like his little drops of planetary lake water, each of us is carrying around our own microscopic universe.

The first real efforts to map out this population of microbes living "with" humans have just begun. Modern tools of genetic analysis let us make a biological census of any environment by gauging the diversity of certain common genes—stretches of DNA code that are directly responsible for some key biological function. This enables us to survey not only the soil or the ocean but the landscape of our own flesh and blood, and it's worth considering some of the results, because they're leading us to a fresh perspective on our microcosmic, and cosmic, standing.

Take, for example, what lives inside our lungs. Current estimates are that more than two thousand individual microbes live in every square centimeter of the convoluted interior of human airways, and that these organisms belong to at least 120 distinct species. The total inner surface area of a healthy pair of adult lungs adds up to about 70 square meters, or 750 square feet—almost a third the size of a regulation tennis court. Thus these 120-plus species number close to 1.5 *billion* individuals (and that may be a severe underestimate).

Until recently it was assumed that our lungs were essentially sterile. Mucus or tissue samples taken from people and used to try to culture bacteria didn't really yield anything. We now realize that this is because these microscopic inhabitants simply won't grow outside of the human lung environment. They require this particular niche to survive.

It's enough to make one feel a little peculiar, but there's more! To get the right perspective it helps to remind ourselves about our human genetic code, the information contained within the long strands of DNA molecules packaged into the nucleus in each of our eukaryotic cells. Altogether it is about 3 billion characters long. This human genome contains some 20,000 to 25,000 distinct genes that code for proteins—which sounds like a lot, until we take a look at another of our microbial habitats: the rich jungle living inside our digestive systems.

In 2010 a group of scientists based in Europe announced the result of a genetic census of the human stomach and gut microbial fauna. From more than 1,000 distinct species of organisms, they found about 3.3 *million* genes—an extraordinary number that is 150 times larger than our purely human set. Furthermore, by focusing on less than 10 percent of the species of bacteria in the human gut microbiome, biologists have discovered within their genes some 30,000 codes for previously unknown proteins. It seems that these human-based critters have a remarkably rich and diversified toolbox of biological machinery.

That's a good thing, too, because the more we study this microbiome, the more we realize how we rely on it. Some dependencies are relatively straightforward. For example, *Bacteroides thetaiotaomicron*, a bacterium found in many animal digestive systems, can break complex carbohydrates into much simpler sugars and other molecules that the host organism can utilize. Our genetic makeup lacks the codes to make enzymes that can cope with these carbohydrates. By contrast, this bacterium can produce a staggering 260 enzymatic varieties, turning us into veritable herbivores, able to digest and extract what we need from all manner of fruits and vegetables.

Other microbial dependencies can be subtle but profound,

ranging from the ways in which we are triggered to feel full or hungry, to convoluted chemical interactions that help stabilize and control our most basic immune-system responses. A number of biologists have proposed that the human microbiome be classified as another major organ in our bodies. Some have also suggested that it does not make sense to separate out our genes from those of the microbes—they should all be considered together. It's beginning to look like they might be right. And the microbiome has another characteristic that pushes this concept to a new level: the exceedingly personal nature of our single-celled companions.

As scientists have studied the diversity of species in our bodies, subjecting them to the probes of modern genetic analysis, they have found that people come in different bacterial flavors. This flavoring varies with what piece of the human body you're talking about: guts, lungs, mouth, hands, or other nooks and crannies.

For example, we think that every human gut microbiome falls into one of three major types, or enterotypes in the terminology of microbiologists. There is no apparent relationship to a person's gender, age, or size, although we don't yet know what variations may be due to location on Earth.

This discovery means that we each carry an unseen microbial label that must relate to our personal characteristics, from the ways we digest and assimilate food to our overall body chemistry. Because our gut bacteria play such a vital role—providing enzymes that help synthesize vitamins, for example—the particular populations that we carry around must play into the most fundamental mechanisms of survival and natural selection. If I harbor one microbial flavor, it may make me better or worse at facing certain environmental pressures than my friend who may harbor one of the other enterotypes of microscopic fauna.

It's clear why we end up colonized by these microbes (we couldn't function without them), but we don't yet know exactly how and when in our life cycle it happens. Much does appear to happen when we're infants, through contact with other people and our surroundings. There is also growing evidence that we are already inoculated in the womb, and further occupied by maternal and environmental microbiomes during birth and nursing. But what really determines the flavor we end up with as adults—or even the extent to which it can change with time—is still a mystery.

More surprises undoubtedly await us as we learn more about this biological universe within. Researchers now speculate that even aspects of our personalities, and our traits inclining us toward friendliness or aggression, are chemically influenced by the particular bacterial composition of our microbiomes—almost a shadowy "microbial soul."

We are not quite the individuals we once thought we were, and that is profound. It means that we have far more in common with the planet beneath our feet than we might ever have expected. Just as Earth's environment has been molded and engineered by microbial life across nearly 4 billion years, our own functioning, our own evolution, has also been directly governed by a part of the core planetary gene set operating within our own cells and out of our bacterial passengers. There appears to be little separating us from the laws of the microcosm.

❧

What we've learned about the nature of life on Earth tells us many things about our significance, as well as that of all life on this planet. As a distinct species of eukaryotic organism we represent a particular instance of life's diversity, yet this doesn't necessarily give us any special status in the microcosm. In many ways it might be better to reconsider the entire hierarchy of life on Earth by placing the microbes at the top rather than the bottom. After all, how we've cataloged and organized terrestrial life over the centuries is a function of our own stage of discovery and understanding. The modern mapping of the "tree of life" based on genetic analysis already rearranges the hierarchy in many respects.

Biologically, the most "important" organisms on the planet, the ones that really determine the specific nature and history of life's ensemble, are arguably not those that appear the most "complex." The most influential members of the community of life are not the multicellular animals and plants, but rather the organisms that to-day use these larger constructions as mobile or expendable environments. They're the bacteria and archaea that have profoundly reworked the chemical and physical environment of the planet over billions of years.

Take humans, for example. For the bacteria and archaea that call us home, we represent a versatile and useful system. Our physiology drives us to find food, and we are even driven to find food that appeals to us because it provides nutrients our bacterial passengers need. Luckily, our anatomical structure and brains also provide us with the means to arrange to bring food to ourselves. We can hunt, and we can plant crops. Given time, we can establish global networks of food production and transport that assemble a smorgasbord of nourishment and even store it in protective buildings so that we—and our passengers—shall not want.

Our analytic brains also come up with sophisticated mechanisms to support the continued functioning of not only our individual selves, but entire groups and colonies. From clothing to heating, shelter, and even medicine and pharmaceuticals, we have developed the means both to increase our short-term chances of survival and to hugely extend our paltry individual "use-by" dates. But what has really done this? Our own needs, or the needs of our microbial overlords pulling on the reins of natural selection?

It's an interesting exercise to consider how we might be characterized by some external, uninterested party. It's not hard to see how our entire species might be described as a population of drones, at the beck and call of single-celled life. Well-engineered drones, to be sure. One of the costs of producing a versatile mobile platform is that it needs some autonomy to respond to its circumstances. We ourselves build sophisticated robots to perform repetitive but extremely precise mechanical tasks. We're also giving them the rudiments of decision making in order to make them work more efficiently at serving us.

For example, a modern car is packed with computer systems and algorithms that allow the machine to make "choices" about its circumstances in order to optimize its use of resources and to maintain the safety of its occupants. The rovers that we've sent to Mars have a limited ability to determine the quality of their track across the planet's surface. It's a fail-safe mechanism to circumvent the fact that a round trip for signals from Earth can take twenty minutes or more—an impossibly long time if you find yourself teetering on the edge of disaster. This pattern of engineering optimization is arguably the same that we find in our own biology.

The notion that humans are just drones for microbes runs headlong into all sorts of issues, of course—not least contemporary evolutionary theory and the mechanisms of developmental biology, and obviously our intimate sense of identity. It's simply a question, not a serious hypothesis—a way to suggest an alternative conceptualization of our significance here on Earth that is also consistent with the vast panorama of planetary microbes and core gene sets. There is no need to suggest that the microbial beneficiaries of our existence (or that of any multicellular species) are actively planning or directing our evolutionary behavior. Rather, this arrangement can simply emerge through a tightly integrated symbiotic or endosymbiotic relationship, whereby mutual benefit helps drive change.

There is a plausible connection between these ideas and our questions of cosmic significance and cosmic uniqueness. I think that we are seeing here a set of chemical and biological constraints and opportunities for complex-celled life that could represent another universal rule. Microbes may always need to be in charge of larger organisms. That's a further tweak to the basic recipe for life's survival that we've been deciphering. We can add it to the core planetary gene set, and to the way the processes of metabolism are assembled out of a ubiquitous carbon chemistry that arises directly from the deeper laws of nature.

This means that while our personal biology may have details that are indeed unique, the fact that the core planetary gene set on Earth evolved creatures like us isn't necessarily surprising, and it might not be surprising anywhere else in the universe, either. This is an important idea, but we should hedge our bets a little longer, because many other mysteries remain to be solved.

For example, we're just part of an ocean of biochemical

components that washes through the outer layers and the atmosphere of Earth. This great wash of molecular possibilities harbors many other, poorly understood realms, among them the abyss of the viruses and the bizarre tangles of prions—misfolded proteins, perhaps the buffer overflow of biochemical errors or spare parts. These agents participate in the biological equivalent (though many orders of magnitude larger) of the subatomic, or quantum, world. Large molecules and snippets of genetic code are transported, exchanged, inserted, and removed. This machinery is still obscure to us, but it must be a vital actor in our planet's history.

So, although it's reasonable to regard humans as life-forms floating in this biological cosmos, does that really mean that we're not significant on Earth?

There may be some hints to an answer sitting right between our ears. When it comes to evaluating our own status, there is one other factor that rears over all talk of the universal or parochial nature of life. And that is the question of intelligence.

No matter how much we may love our dogs and cats, or look with sympathy at chimpanzees, elephants, or dolphins, it is abundantly clear that *in toto* there is something different about humans. The intricacy of our brains, the social structures we form, our cognitive skills—from languages to the ingenious problem-solving and reasoning that we habitually employ—are all far off on one end of the spectrum of such things for life on this planet. Yes, chimps, rats, and even crows may come close to having some of our sophistication in thought, and share a large percentage of our genome. A glance at the incredible social order of creatures like ants is humbling, as is also true for the multitude of forms of communication used by life across the planet. But to have all the pieces in one organism, one creature—that seems to be a singular occurrence here on Earth.

The idea that we stand alone in 4 billion years of evolution on this planet is a thorny item to bring to our quest to understand our cosmic significance. The questions it raises spill off in different directions. For example, how might the broader traits of intelligence propel evolution on other worlds? Even here on Earth, the humble and wonderful octopus, a member of the cephalopod family, has a very different nervous system than any vertebrate animal like us, yet it can manipulate objects with incredible dexterity, and opportunistically uses materials such as coconut shells much as we use tools. Is there a planet of the cephalopods out there somewhere?

Another question has to do with the rarity of our specific kind of intelligence. A huge number of scientists have at one time or another argued that some unique aspect of our mental or physical composition is responsible for our "rise" as a species. The human hand, our knack with language, our omnivorous digestive abilities, our social tendencies: qualities such as these and more have been pointed to as being responsible for our survival and key to the evolution of our type of intelligence. But perhaps none of these attributes were inevitable results of natural selection. Maybe they were sheer dumb luck—after all, brains like ours may have arisen only once in nearly 4 billion years of life on Earth. That hardly suggests a great evolutionary strategy.

All of these observations can reinforce the idea that Earth is a rare place, an unlikely world where our existence results from a chain of lucky steps and mishaps. Perhaps. While this view *may* be correct, on the other hand it may be distorted by our often horrible intuition for certain types of statistical conditions, which I will tease apart soon enough.

Fueling some of this debate has been an explosion of information in recent years about what makes us human. Both paleontological discoveries and genetic analyses are leading to a fantastically interesting picture of where we come from and what we represent in evolutionary terms. Some of these newly revealed aspects of our existence tell us we're lucky to be here, but others indicate that the constant exploration of survival strategies by natural selection and evolution is always finding novel successes that could help explain the likes of us.

For instance, genetic studies suggest that between 123,000 and 195,000 years ago the population of biologically modern humans declined dramatically, from more than ten thousand to just a few hundred. We don't know exactly what went wrong, but it's likely that climate change was partly to blame. A long glacial stage kicked in over this period that would have profoundly altered the planetwide distribution of vegetation, animals, and temperate climate zones. Deserts appeared where there had been none, and readily inhabit-able regions may have diminished in area.

Somehow, though, a few humans survived, perhaps living off rich pickings in coastal areas below the equator, where the remains of countless shellfish dinners have been discovered. All of us alive today come from this tiny group of people living somewhere in cen-tral or southern Africa more than a hundred thousand years ago.

It doesn't take a great deal of imagination to realize that modern humans could well have ended their run then and there. Disease, or further worsening of the climate, could have easily polished off those few hundred individuals. Sheer chance may have saved our species close to its outset, but intelligence may also have helped us escape the clutches of extinction.

We weren't the only ones to survive that period. At least one other tool-making biped was walking around the Earth at the same time as modern humans. By about 600,000 years ago, the species now known as Neanderthals had, we think, migrated from Africa into Europe. The Neanderthals were like us in many general ways, but also very different. They were another flavor of upright ape-related creature, which we think evolved from an earlier species—a common ancestor. They were not stupid; they made stone and bone tools, and they were social.

Yet Neanderthals drifted to extinction about 28,000 years ago. Exactly why, we don't know. Further climate changes, and even competition from modern humans, could have played a role. But remarkably, a part of them is still with us: the genetic code of people from Eurasia contains as much as 1 to 4 percent of the Neanderthal code. We know this because we've been able to decrypt large frac-tions of the genetic sequence of Neanderthal remains, an amazing and spooky piece of detective work. It means that sometime after the near-extinct population of modern humans managed to travel to and flourish in the planetary north, there was interbreeding with the Neanderthals. And then we survived and they didn't.

In addition to studies like these revealing our sometimes peril-ous history, there have been a number of discoveries about the fun-damental molecular machinery that makes us a distinct species. These findings bring us back to the question of our significance be-cause they include direct insights into what distinguishes our form of intelligence. Genetically, modern humans are only about 1.2 per-cent different from chimpanzees, and by seeking out those differ-ences, we can look for specific functions coded by those genes that may be uniquely human. Some of the sequences of DNA that differ most between humans and chimpanzees are indeed directly related to critical aspects of what sets us apart from other creatures.

For example, the sequence known as HAR1 (human acceler-ated region 1) is active in the brain and may relate to cerebral cortex development. We think that another sequence, HAR2, is involved with our embryonic development and the construction of our wrists and thumbs—hallmarks of our ability to manipulate material and use tools. The sequence known as LCT is associated with the adult ability to tolerate lactose—to consume dairy products. Remarkably, studies indicate that this sequence is very recent in evolutionary terms, and indeed, only about a third of humans worldwide—but 80 percent of people of European descent—have it. Although most mammalian species alive today can digest milk sugar as infants, they lose that ability as adults. Some 9,000 years ago that changed for a human group, with the rise of a version of the LCT sequence that continues to produce the necessary digestive enzymes in adult-hood. Keeping domesticated animals would take on a whole new set of advantages from then on.

There are other important sequences that are linked to human versatility. AMY1 is involved in the production of an enzyme that lets us digest more starch than many species. ASPM is a DNA se-quence that affects brain size. And perhaps the most unnerving sequence of all is FOXP2—forkhead box protein 2, which research suggests is involved in the way our faces and mouths move to pro-duce the multilayered sounds and rhythms of speech. While we can find similar sequences in most other mammals, the version in humans is quite specific and different from, for example, that of chimpanzees. Without speech, our incredible social structures and our ability to pass on information and share experiences would obviously be far different. This piece of DNA, a mere 2,285 nucleic acid bases long, may be critical in making us human.

The genetic differences between us and our closest relatives, the chimpanzees, are not all good news. There is evidence in our genes of ancient battles with retroviruses, structures that replicate by injecting their own genetic material into their host's. In some cases we eventually came out of these fights equipped with code that makes us much better than other primates at resisting these insidious pathogens. But these same genes make us more susceptible today to something like the HIV retrovirus than our ape cousins. Like our struggles to survive more than a hundred thousand years ago, our genetic history has not been without peril or chance.

As we continue to break down our personal functioning into these pieces of molecular machinery, it remains an enormous challenge to relate these discoveries to the riddle of how we evolved. It's certainly clear that being intelligent has been part of a great survival strategy thus far. Yet there's so much more to our history of pushing and shoving through the barriers of natural selection. Abilities to digest certain foods, grasp objects, adapt to certain ranges of temperature and of wetness and dryness—all these come into play. And the external driving forces of climate and other species' successes and failures have all been major influences.

But through it all, as special as we are, our story also parallels the story of any other complex-celled life-form on this planet. Each has its own special genes, and its own unlucky and lucky evolutionary shifts. The engineering of biochemistry does all this. It's a nesting of machines inside other machines, all the way down to the most fundamental and basic nature of atoms, which blend into the quantum and subatomic realm. The great experimenter that is evolution throws a billion options out there, in a colossal network of interactions and variations. The pattern for this network, embodied in our core planetary gene set, may be more universal than parochial. It may also be that multicellular, intelligent life is just as likely to spring from such networks as anything else, given the right opportunities.

So is our brand of intelligence unique in kind, or especially outstanding, or especially unlikely? Seen from the receiving end, it can certainly seem to be a bit of all three. But this not only conflicts with the basic tenets of a Copernican worldview—demoting us to cosmic mediocrity—it's also not currently testable. In fact, it won't be until we know better how to evaluate the importance for evolving intelligence of all those branching pathways of life on this planet, and, most critically, find out whether those that are make-or-break can happen elsewhere. The biological universe therefore brings us face-to-face with the greatest challenge of all in our quest to learn our cosmic significance.

Are we alone?

6

HUNTERS OF THE COSMIC PLAIN

If I had to name two traits that accurately and optimistically sum up the human species, I'd put my money on imagination and restlessness. The signs are everywhere. Take, for example, the ways in which we've expressed our fascination and vexation over our cosmic status. Artifacts and expressive records of our observations and imaginings from a thousand, five thousand, even twenty thousand years ago reveal intense cogitation. Although anthropologists continue to debate the motives behind the most ancient cave paintings and sculptural forms, one of the most plausible theories to me is that they reflect an effort by early modern humans to analyze their universe of animals, landscapes, and rituals. It might be tempting to believe that these illustrations or objects were simply idle doodling and fiddling to while away a boring winter, but even if that's so, I can't help but feel that there was something conscious and deliberate going on—perhaps a sorting and sifting of facts and observations that had not quite coalesced into a rational picture of the world. And this behavior didn't happen once; it continued from generation to generation. Some of the more abstract of these ancient images and figurines dwell on strange representations of human-animal hybrids, Earth mothers (possibly deities), and monsters. It is the stuff of feverish dreams. Here's the human mind working desperately, trying to fill gaping voids in knowledge, and to understand the Meaning of Life. If we need to invoke unseen beings and forces to make sense of it all, then so be it.

The same has been true of our struggles to map the relationship of the heavens to the Earth, the Sun, and the Moon, often connecting the planets and constellations to gods and fantastical creatures in an attempt to provide explanations for the patterns we saw. The nature of time has also been perplexing—to our ancestors as they studied their surroundings, and still to humans of the twenty-first century as we theorize about the nature of the universe. On all physical scales the cosmos embraces change, clearly moving along to leave the old and weak behind, and to weather rock and rot and decay the husks of living things. We also observe and record great regimes of repeating seasonal change, lunar cycles, and the slow pulse of climatic variations. What goes around comes around. Witnessing the cycles of biological life, we have extrapolated to invent the notion of endless cosmic repetition and rebirth, concepts that span human cultures and generations.

This creative hubbub of drawing, mapping, and timekeeping has at its core a thirst for cosmic clarity. And again and again we arrive at the question of whether or not anyone else is "out there," in space or time. Yet one can easily argue that there has never been any data at all on the presence or absence of other life in the cosmos. I don't want to make this sound too depressing, but it's true—which is why we're lucky we've discovered beer and chocolate to console ourselves.

This rather dismal isolation and ignorance hasn't stopped us from making grandiose claims through the centuries and millennia. One of the most entertaining historical flavors of speculation about the nature of life beyond the Earth is the notion of the plurality of worlds. We've bumped into this concept before; it's had a long gestation, going back to the great philosophers.

Some of the ancient Greeks, the atomist Democritus for one, believed that if the underlying nature of reality was granular—consisting of indivisible atoms and voids—it implied the existence of an endless variety of objects such as planets, suns, and moons. This wasn't necessarily meant to suggest that an infinite number of worlds were all "out there" in the tangible universe, which was observed in an extremely limited fashion, but that they would exist *somewhere*. Such an expansive vision of our cosmos led certain followers of this philosophical school, such as the thinker Metrodorus in the fourth century B.C., to argue that in an infinite realm it would be awfully strange and improbable if there were only one place like the Earth. But when Plato and his cohorts (Aristotle among them) came along a few decades later, they managed to squash this notion, positing that the Earth was both central and unique in creation.

Despite these setbacks, the idea that there are other worlds has persisted in the human imagination, as I've touched on earlier. Long after ancient Greece washed its hands of the notion in the third century B.C., the plurality argument reared its head again—first during the medieval period in the Middle East, and then in the late 1500s in Western Europe with the likes of Giordano Bruno, who so wholeheartedly embraced the implications of Copernicus's cosmology. Indeed, the Copernican decentralization of the cosmos reopened the floodgates to the concept, and it gathered momentum in the following centuries. The notion of plurality was often inseparable from the notion that these worlds were also inhabited. Plurality of worlds meant plurality of life. In many respects this thinking was a perfectly logical continuation of the Copernican model that the Earth was not central to the universe, and was not unusual.

In the late 1700s the brilliant William Herschel, a German-born but Anglicized astronomer who discovered the planet Uranus, became enamored of the argument that there was life on other planets. It seemed more reasonable to him, as it did to many other scientists, that other worlds should be full of people and creatures rather than barren and empty. This logic also allowed for the comforting possibility that the same religious and social orders existed everywhere— a clever way to be both decentralized à la Copernicus and still cosmically important by virtue of our participation in a grander scheme. After all, if we drank afternoon tea and went to church on Sunday in bucolic England, *surely* the same thing would be happening on Mars.

Some of this thinking took even more inventive turns. Herschel mused that the Moon was inhabited by intelligent beings, and went so far as to declare that in his telescopic observations he felt sure he was seeing something akin to a forest on one of the lunar maria, or plains: "My attention was chiefly directed to *Mare humorum*, and this I now believe to be a forest, this word being also taken in its proper extended signification as consisting of such large *growing substances* . . . And I suppose that the borders of forests, to be visible, would require Trees at least 4, 5 or 6 times the height of ours. But the thought of Forests or Lawns and Pastures still remains exceedingly probable with me . . ."

He even felt that the Sun must harbor a hot atmosphere shielding a cool surface, glimpsed through sunspots that he thought, incorrectly, were gaps in this gas. Naturally there had to be inhabitants. As Herschel explained in 1794, "The sun . . . appears to be nothing else than a very eminent, large, and lucid planet . . . [which] leads us to suppose that it is most probably also inhabited, like the rest of the planets, by beings whose organs are adapted to the peculiar circumstances of that vast globe."

Herschel's ideas about life on the Moon or the Sun were certainly not mainstream, but they weren't entirely on the fringe, either. Even the famous and brilliant French mathematical physicist Pierre-Simon Laplace discussed the possibility of life on the other worlds of our solar system. But it was a little later, in the 1830s, that a scientifically minded Scottish minister and would-be astronomer by the name of Thomas Dick made some of the most extraordinary efforts to quantify the number of beings elsewhere in the universe. His first step was to assume that the population density of the United Kingdom at the time was representative of the density of beings on any other planet or asteroid—a startlingly mad thing to do, at least to our modern sensibilities.

On this basis he went on to estimate that the planet Venus held more than 50 billion individuals, Mars had 15 billion, and Jupiter a whopping 7 trillion. In a wild bit of speculation he even suggested that Saturn's rings held something like 8 trillion inhabitants—just in the rings! Having completed all this enthusiastic extrapolation, he pegged the solar system's net population of living beings at about 22 trillion—not counting the Sun, which he pointed out could alone harbor thirty-one times *more* creatures. And he didn't see fit to stop there. He also came up with an estimate of the total number of *planets* in the universe being in excess of 2 billion, all of which could be inhabited with the same density of beings as those strolling around the sceptered isle in the 1830s. Ironically, this count of planets we now know to be woefully low, but in fairness, no one at the time knew the true scale and extent of the universe.

The motivations behind Dick's projections (which were at the absolute extreme end of pluralist arguments) are still important to consider, because lots of serious scientists felt a kinship to them. There was no way to obtain incontrovertible proof that other worlds were or were not inhabited, and for many people it was simply easier to assume that they were. Even with the best telescopes of the time, it was unlikely that anyone would be able to genuinely confirm or disprove signs of life elsewhere. No images had the necessary resolution to help astronomers see the comings and goings of creatures on another planet.

Without evidence one way or the other apart from the Earth, an abundance of life on all celestial bodies could be seen as a natural part of planetary existence, like another layer of material that complemented the rocks and soil. If no other worlds were inhabited, then we would have to find a good reason why. The logic of this reasoning is hard to argue with. Once again, anything that sets the Earth apart from other places is awkward if you fully embrace a Copernican worldview, which was the scientific consensus at that time. It was better to populate the cosmos than make Earth unique.

But time has passed, telescopes have improved enormously, and our appreciation of the actual properties of life has changed irrevocably with the realization that organisms are not static entities. They are products of an ongoing and complex process of evolution and natural selection. Somewhere along this line of scientific thought, planets ceased to automatically equal life. Organisms don't just plop down en masse. We recognize now that life might—or might not—be able to occur in certain places. The most extreme ideas of the plurality of inhabited worlds have faded from view, and today are firmly consigned to the scrap heap. Our exploration of the solar system has quenched the notion of complex life on the Moon, Venus, and other of our neighbor worlds. And even though we now know there are an overwhelming number of other planets in the universe, we also know that organisms like ourselves cannot occupy them *all*, because conditions on many of those worlds won't allow it.

But we are left in a curious intellectual spot, because the universe is obviously a big place. Within our observable cosmic horizon—the distance over which light has managed to travel in the 13.8 billion years since the Big Bang—are several hundred billion galaxies and potentially more than a billion trillion stars. And that's just the number that appear to us at any one instant, a mix of objects in a snapshot assembled from countless cosmic moments when distant light set out across space to us. Ask how many stars have *ever* existed in the past 13.8 billion years, and, apart from inducing a headache over the concepts of time and space in a relativistic cosmos, you'll end up having to wave your arms wildly in the air to justify quoting an even bigger number.

This empirical fact is centrally important to our struggles to understand whether or not anyone else is out there. A huge universe motivates a different kind of answer than a tiny one with few suitable places, and it's the kind of answer that we've all heard before, and probably even thought of ourselves. Since the universe is so big, filled as it is with a billion trillion stars, there surely *has* to be life somewhere else.

But does the gaping enormity of the visible universe really lead to the inevitable conclusion that there must be someone else out there? The question of "aloneness" contains other hidden layers, too. In particular, much like the pluralists of old, when we ask that question we're usually wondering whether or not there are any other creatures *like us* in the universe: thinking, contemplating, technological or philosophical beings, with ideas, beliefs, art and poetry, and, of course, science. And, as with so many phenomena in our world that seem obvious, we would do well to step back for a moment and take a careful look at the details. In this case, a vital issue is whether or not we can tackle the implications of a massive universe with a mathematically rigorous analysis. Can we formulate a properly scientific response, one that moves us beyond the imaginings of pluralists or plain old knee-jerk optimism?

We can. And formulating such a response starts in the unlikely world of probability theory.

❧

What's funny about reading accounts of the life of Thomas Bayes is that quite a few begin by stating that he "was *probably* born in 1701 . . ." In fact, the historical record of his life and even his mathematical works are rife with uncertainty, due to a relative paucity of documentation and his own apparent reluctance to publish all of his scientific work (a wonderful irony in the light of what he remains best known for). We do know that Bayes was the son of a Presbyterian minister in England, from a family made wealthy by the steel cutlery industry, and that he studied mathematics and theology at the University of Edinburgh before becoming a minister in the late 1720s.

Around this time he published some theological work, but his scientific interests truly flourished. During this period Newton's theory of differential calculus—more generally known at the time as a "method of fluxions"—was not universally accepted. In a nutshell, Newton's theory showed how to describe the rate of change of any mathematical function (from the arc of a cannonball to the slope of a curving surface) as its parameters varied, and was deeply connected to the notion of infinitesimal divisions of these functions. Newton used the word "fluxions" to mean the action of flow or change.

Bayes's only other officially published research during his lifetime was an attempt to support Newton's theory by providing more-rigorous proofs of the mathematical properties of fluxions. It may not have been that exciting, but work like this was good enough to secure Bayes a coveted fellowship in the Royal Society, and to encourage him to carry on his mathematical pursuits.

Later in life Bayes became interested in the theory of probability, a field of study that had been emerging among mathematicians over the previous century. It was a provocative area of work, not least because it dealt with issues that could be discomforting to those with a strong belief in a higher power. Scientists were beginning to appreciate that chance and uncertainty might really be just that: random events that occur in the universe without plan or purpose. This was profound: it signaled the start of a deep shift in our conceptual model of nature.

But it wasn't until the philosopher and preacher Richard Price sifted through his friend Bayes's papers and notes after Bayes died in 1761 that it was discovered he had made significant progress at solving some of the thorniest problems central to this subject of mathematical "chances." Price was instrumental in gathering up this work and getting it published by the Royal Society two years after Bayes's death. As a result, Bayes is now best, and famously, remembered for tackling a problem known at the time as "inverse probability." We don't use this term much these days, referring instead to posterior probability or inference. Over the following decades and centuries, scientists such as Pierre-Simon Laplace independently discovered and built on these concepts to extend them—and they now infuse all of modern science. But it's Bayes's name that stands out, enshrined in what has become "Bayes' theorem," the distillation of his last and greatest work on probability. The theorem can be written as a simple formula. It provides a mathematical expression for the probability that a model or hypothesis is correct, given a set of actual observations. Most important, it boils down to a way of looking at the world that allows us to evaluate our *confidence* in a theory or a prediction.

The essence of this framework can be explained with a little allegorical tale that Thomas Bayes's friend Price sketched out as a note in the posthumous publication of Bayes's work. I'll paraphrase: Let's imagine that there's a rather mathematical, but unfortunately naïve, rooster that's just hatched. On this, his first day in the world, he's surprised that the sun moves across the sky and eventually vanishes out of sight. The rooster wonders whether he'll see that bright disk again. Being analytically inclined (for a baby chicken), he formulates a simple hypothesis that there is an equal probability of the sun reappearing or not reappearing—his confidence is split 1:1, or 50 percent.

Of course, some hours later the Sun does indeed rise above the horizon. It moves across the sky again and then disappears. The rooster decides to update his expectations (or, his confidence). He's now seen two sunrises, but there's still a possibility that these won't repeat, so his odds become 2 to 1 (66.7 percent) on there being a third sunrise. With this next sunrise he updates his confidence again, to 3-to-1 odds (75 percent) in favor of the sun returning the following day. And so it goes on, with each passing day increasing the rooster's confidence ever closer to 100 percent that the sun will indeed rise again. By the hundredth sunrise he's about 99 percent sure, and decides he could probably get away with crowing just before dawn, much to everyone else's dismay.

The rooster's analysis is a basic example, but it's at the core of the Bayesian approach to data and theory. The outcome of experiments, and new observations and data, can all modify one's confidence in a hypothesis, to evaluate its probability of being accurate. But it wasn't always clear to scientists that the quantification of uncertainty was an approach that made sense. In fact, it wasn't always considered reasonable to think about experiments or observations this way at all—to make the world a place of probabilities or "confidence" in what was true or not. It took a long time for these concepts to sink in. A scientist as eminent as Herschel (who was considering the existence of life on other planets just a few decades later) obviously didn't draw his conclusions this way. So we owe a vital debt to Bayes and others in the 1700s who were trying to figure out how to take *uncertainties* and turn them into *probabilities*, as our allegorical rooster did.

We can see how Bayes himself was trying to come to grips with this challenge by the example he used to clarify his mathematical formulation for his readers. He imagined carrying out an experiment, most often illustrated by using a billiards table as a model—although in truth no one knows if Bayes actually meant a billiards table or just any old table. But for the sake of the story let's stick with billiards.

Suppose you carelessly roll a red ball across an empty billiards table so that it may come to rest anywhere. You leave the red ball where it is and then roll a white ball in the same direction across the table a number of times, adding up how often it comes to rest on the far side of the red ball. Now, Bayes used the pattern of balls on this imaginary table to derive a mathematical solution to a simple question: Given what happened with the balls you already rolled, could you predict the chances (or probability) of the outcome of rolling one more? In other words, what are the odds of your next roll ending up on one side or the other of the red ball? Bayes showed that you could figure out that probability. Critically, like our rooster with the sunrise, the more balls you rolled, the better and better your confidence about the odds of the outcome of the next roll.

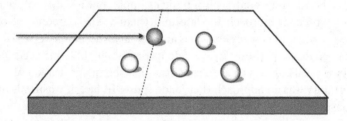

The billiards thought experiment was a simple one, but it speaks to just how fundamental the question was for mathematics in the 1700s. No one had figured out how to construct the necessary mathematics in any detail before, and the conceptual pieces involved in handling uncertainty were staggeringly new for everyone. Bayes was moving toward a formulation—his theorem—that could be used to calculate how an individual "believed" in a hypothesis in the face of evidence—how a likelihood or confidence could be assigned to a proposition's being correct.

To understand the composition of that theorem, and to begin to see how we can apply it to the question of life in the universe, I like to use a slightly more colorful example than sunrises or billiard balls. Let's suppose I have a curious hypothesis that Cheshire cats make up 20 percent of the population of all cats. In order to test this hypothesis I need to go out and find some cats, and I need to try to identify and count the Cheshires and the non-Cheshires. This is a challenge not unlike that of looking for signs of other life in the cosmos, finding inhabited or uninhabited planets.

Of course, counting cats is easier said than done. I'm shooting in the dark, with no preexisting information to guide me. For one thing, unless I'm willing to capture and evaluate an enormous number of cats, there will always be significant random spread in my results. If I bag ten cats off the streets and identify two of them as being Cheshire, I can't say *for sure* that this confirms my hypothesis of a 20 percent fraction of Cheshires, because there will be a large margin of error due to the random sampling of a small number of cats.

So my theory about Cheshire cats has to be a bit more sophisticated, and it has to include some expectations about the spread (or variation) of randomly selected groups of cats. In effect, it has to be able to predict that margin of error—to tell me what I would expect my measurements to look like if my hypothesis is correct.

In addition to the complications posed by random sampling, there is the issue of more systematic bias to my survey. Maybe Cheshire cats are typically overweight and slow, so it's easier for me to catch them and count more of them. Perhaps my hypothesis about Cheshire cats is completely wrong from the outset (a distinct possibility, since they are prone to bouts of invisibility). But I might delude myself into thinking it was correct if by chance I just happened to find the right number of what I thought were smiling Cheshire cats in my random collection.

So the probability of my Cheshire cat hypothesis being correct is itself equal to a mathematical combination of some other, related probabilities. The first of these is the probability of the data or measurements *given* the hypothesis. That may sound a little odd, but it means that if a model or hypothesis were correct, then you would expect your cat counts to follow a certain pattern. For example, I would be able to assign a specific probability that I'll count 1, 2, or 3, or any number of Cheshire cats in my random catch of 10 cats.

The next piece is what's known as the *posterior* probability, which is the thing we're really after for the cats, as well as for the question of life in the universe. The posterior probability is the more intuitive reverse of the above. It's the probability of the hypothesis being correct in light of the measurements or evidence. In other words, this probability would tell us how likely it is for my theory about cats to be correct, or how likely it is for there to be other life in the universe, given only the observation of life here on Earth. It's also the same measure of *confidence* that we saw with sunrises and billiard balls.

And finally, in my cat-based description of Bayes's formula is a factor that is our hypothesis itself, which is called a *prior* probability. In this case it is the probability that any cat is Cheshire, which we think is 20 percent, or 0.2. Of course, we don't *know* whether 20 percent is correct or not; it's the very quantity we'd like to confirm, much like the probability that any single planet could produce life. The interesting point is that by assigning this probability we are implicitly assuming that the very idea itself, the existence of Cheshire cats, is correct. That kind of assumption is dangerous, because we could incorrectly give too much weight to crazy hypotheses. So unless we're sure of ourselves, the best thing to do is to evaluate many possible "priors," and cross our fingers that the data we have will sort out the winning and losing hypotheses by their relative probability.

The formulation of Bayes' theorem I've outlined here also assumes that whatever data we acquire is accurate—that there is no type of "false positive" or "false negative." So if I pick up a cat in my little feline survey I assume that if I identify it as a Cheshire cat, it really is. That's an important caveat. In the medical world, for example, false positives or negatives abound. In these cases the Bayes formula gets a little tweak to include the probabilities of misdiagnosis, and of faulty chemical tests. If you're trying to evaluate the likelihood of a particular illness, or an emerging epidemic, the accuracy of your data and the priors you choose are critical factors.

So Bayes' theorem lets us evaluate the relationship between what we can observe and measure and our hypotheses, or mathematical models. In principle, it should let us assign an absolute probability, a confidence, that a hypothesis is an accurate description of a phenomenon in nature. But there are some sticky problems that can make that calculation hard to swallow. We may not know exactly what the prior is, or whether our hypothesis is remotely correct at all. And our measurements may have imperfections due to random sampling or unanticipated errors—a real problem in my example, because Cheshire cats are entirely fictional. So the probabilities, the confidences, we calculate may have small absolute values that don't help us in our decisions.

Luckily, Bayes's idea is far more powerful. There's a way past these apparent obstacles, a clever sleight of hand that scientists have learned to apply in their daily work—whether it's tracking cats or calculating the structure of the cosmos. The simple fact is that we usually don't care what the *absolute* values of the various probabilities are. What we really care about is whether one model or hypothesis is "better," or more probable, than another. So we can start out by assuming that all hypotheses are equally likely to be right. What is truly important is finding out which hypothesis is best matched by the data or measurements we have—which one wins. They could all be wrong, yes, but we simply want to know which is *least* wrong. We can do this by flipping Bayes's formula around. We end up simply evaluating the probability, or confidence, that our measurements could be produced by a given hypothesis *compared with other ones*. This quick trick produces an incredibly powerful tool.

What I might do to apply this trick to the case of Cheshire cats is to test a variety of methods for identifying Cheshires, such as calculating how much a feline weighs, or whether it can smile. If 20 percent of cats really are Cheshires, then both accurate and less-accurate methods ought to show some consistency, with varying relative probabilities. The Bayesian approach lets me combine all of these to construct a measure of my overall confidence in the hypothesis compared to alternatives.

And what if none of my identification methods reveals a consistent percentage of these animals, resulting in a low overall confidence? In this case I'd have to consider that either the details of my prior hypothesis were wrong, or there might be no such thing as Cheshire cats. In some respects this is a simple mathematical concept, but it's astonishing how far the application can go. For many scientists, its efficacy at approximating the shape of reality is evidence that Bayesian inference is about as close to "how nature does it" as we can get: it seems to accurately capture the probabilistic outcome of a variety of phenomena that are at their core driven as much by chance as by rules. Except that while nature knows the rules in the situations where we exploit the method, we only have guesses.

Very often this doesn't matter too much. If our guess, our scientific model, is reasonably accurate, the magic of Bayes' theorem tends to cover up the cracks, or at least lets us know how confident we can be in the answers we produce. For some people this is still a disturbing way for humans to draw conclusions about the workings of the universe, as it means that no theory is ever truly *wrong*—it's just not as good as another.

I can vividly remember, as a young graduate student, watching eminent researchers almost come to blows over whether or not we could allow such sloppiness. If Bayesian-style analysis could only provide a *probability* that a particular theory made a good fit to observations of the world, surely we could never entirely trust this approach to acquiring knowledge? Equally, the argument went the other way: surely this was a far more honest and realistic approach to structuring our investigation of the natural world, full of uncertainties and incomplete stories. As with many tasks in human life, though, if something works well enough and provides a decent (albeit imperfect) solution to a problem, it tends to become the de facto solution, and in this case Bayes' theorem easily has the upper hand.

Today, Bayesian inference is a constant presence that has become embedded in our technology and our thinking. It's all around you, far more than you may realize. Almost every piece of clever computer software for photography makes use of Bayesian methods. Face recognition? Yes, at its core is Bayesian probability, making sure that a precious moment of childhood play can be captured in focus. That frustrating ticket you just got for speeding through a traffic light as it changed to red? You can thank Thomas Bayes for that: your car's license plate was recovered from a blurred photo using Bayesian techniques. Autocorrect anticipating (often comically) what you're about to text on your phone? Yes indeed, yet another application of Bayes' theorem—the statistical analysis of word use is generating probabilities for what you're going to type next. The way that automated systems trade stocks and shares, and set the value of commodities and currencies—much of that is done using Bayesian methods to determine probabilities and confidences in outcomes. In the era of Big Data, when companies are gathering information on every piece of our behavior, the same tools of statistical inference and prediction are vital for teasing out useful clues to which brand of soap we like . . . or which brand we might be persuaded to like.

It is Bayes's massively influential legacy in science that is essential to our understanding of what our existence tells us about the likelihood of life elsewhere. Yes, it helps us sequence genetic code, and evaluate test results for cancer markers to estimate how likely it is that you have malignant activity. It lets us scour through pet-abytes of data to find the ethereal signatures of particles and phys-ical laws. But it also helps us tackle the vital question of what our own existence implies for the probability of life's existence else-where in a galaxy of billions of other solar systems. So now that we're thinking about this conundrum like Thomas Bayes, let's see what happens when we try to formulate a mathematical answer to the puzzle of life in the universe.

In 2012, two scientists at Princeton, David Spiegel and Edwin Turner, applied the tools of Bayes' theorem to a more carefully phrased ver-sion of the "Are we alone?" question. They began by asking just what the most reliable facts *are* about life here on Earth. What are the clues we have to go on? This involves stripping away confusing and extraneous information to get at the pure stuff, which boils down to two simple pieces. The first is that some form of life ap-peared quite early in Earth's history, within a few hundred million years of the planet's main assembly. The second bit of reliable infor-mation is that a few billion years later, thinking and question-asking creatures arose to discover this fact. Whittle everything down to the bare essentials and this is the core of what we know about life in the universe at the moment. Pretty sobering.

Next, Spiegel and Turner incorporated this information into a Bayesian formula to ask whether these facts tell us anything about the probability of life arising anywhere else in the universe (abio-genesis). In other words, if life got off to a fast start here on Earth, and a few billion years later evolution produced us, does that imply this is likely to happen elsewhere? As in all Bayesian analyses, there is a tension between the weight—the confidence—we give to the known facts and the weight assigned to our prior assumptions.

So what kinds of assumptions are we making in this case? Spie-gel and Turner realized that by merely writing this formula down, we necessarily make an assumption about the basic probability of rudimentary life appearing on a planet over a period of time. In other words, we assume a value for the *average* number of times abiogenesis will occur in any billion-year period—this is our prior probability.

Now here's the tricky bit. Without a proper Bayesian analysis, we have a tendency to assume that life probably arises readily across the cosmos, or else it wouldn't have appeared so quickly on the surface of a young and cooling planet Earth. But this puts the cart before the horse. It's precisely the same as assigning a value to the average number of times life arises on a planet in any billion-year period when we don't know what this number is!

Spiegel and Turner have called this "prior ignorance," a great way to describe our situation. Taking this into proper account is a little unsettling, because mathematically it turns out that the early presence of life on Earth tells us almost *nothing* about the chances of life occurring elsewhere. Our human instinct to see a reflection of ourselves, our colossal tendency to overstate our importance, gets in the way yet again.

By investigating a variety of mathematical models for this prior ignorance, Spiegel and Turner were able to show that our predictions about life elsewhere are almost entirely a sliding function of what we assume in the first place. Suppose that the (unknown) frequency of abiogenesis on any suitable planet is constant over time. Their Bayesian analysis takes the fact of our existence into account, but says the options for life in our galaxy are still wide open. There could be life everywhere. But it could also be the case that life is a once-every-10-billion-years or every-100-billion-years phenomenon. In other words, we could be the first life in the universe. Tweak the assumptions a little, and all bets are off.

The one example of life on Earth really isn't enough to tell us very much about anything else—we're just like the young rooster witnessing his first sunrise. Yes, life *could* tend to arise quickly on Earth-type planets based on what happened here, but our prior ignorance is such that we cannot rule out the possibility that it doesn't. There's another, subtler side to this analysis, too, and it has to do with the differences between microbial life and human beings. It comes back to the two pieces of information about life on Earth that we start with. We know something about the time span between the emergence of any kind of life on Earth and our own appearance: about 3.5 billion years. What does this mean for the numbers?

Now matters get almost philosophical, because we can ask whether the probability of our presence at this time to observe the universe around us and ask these questions has an impact on the conclusions themselves. In other words, how does the inferred probability of life appearing on any planet change if life requires, as it did on Earth, roughly 3.5 billion years to evolve from microbes to complex organisms capable of computing that probability?

Look at it this way. We might say that a planet *needs* about 3.5 billion years of biological evolution between abiogenesis and the rise of "intelligent" life. If this were correct, a planet of Earth's age where the first organisms didn't appear so quickly would not yet have produced creatures like us. Therefore, it is only natural that we find ourselves on a planet where abiogenesis occurred very early, because on a late-blooming planet we wouldn't be around to make that observation yet!

So the conclusion is that this second piece of information also doesn't tell us *anything* about how likely that first step to life really is on any random planet—for the simple reason that other abiogenesis timings couldn't have happened here anyway (since there wouldn't have been time to produce us to observe the fact). If we carefully step through the mental minefield of Bayesian inference, we come to an unsettling conclusion: we can infer relatively little about the statistics of life in the universe from the history of life on Earth. It may well be that life does usually arise quickly on young, rocky, chemically rich planets. The situation on Earth would then be standard and unremarkable. However, it may also *not* be the norm. The emergence of life could still be a rare phenomenon—we simply have no way of knowing without more information.

The key part of that information is straightforward in essence, but in practice represents one of the greatest scientific challenges of our age. If we could find just *one* example of life originating truly independently of our own lineage, we could dramatically reduce our prior ignorance. The Bayesian analysis even tells us by how much. Instead of a galaxy-wide abiogenesis rate of as little as once every 10 billion or 100 billion years, the minimum would rise to about once every 1 billion years for any single planet. That's a number to get excited by. It need not even be life on an exoplanet. Evidence of a strand of life that had an independent abiogenesis on Earth would immediately improve our knowledge about the cosmic probability of life arising with time.

Equally, independent life on another planet in our solar system would do the trick. Any of these discoveries would increase the probability of life anywhere else in the cosmos, and greatly improve our confidence in estimating this probability. Clearly, our quest for our cosmic significance in the most rigorous and scientific sense will only progress if we go out hunting.

One of the greatest legacies of the Apollo missions to the Moon in the late 1960s and early 1970s was a newfound appreciation for our noble, yet so utterly humble, blue-green marble world hanging in the cavernous blackness of space. But only twenty-one humans flew to the Moon, and only twelve of those ever set foot on its dusty surface. Twelve people total, twelve out of about 110 billion biologically modern human beings who have ever existed. Just to put it in perspective.

We have, however, also done some remarkable traveling in absentia. Ingenious robotic devices have been cast outward from Earth to an impressive array of destinations. Altogether, since the dawning of the space age in the late 1950s we've sent more than 70 missions to the Moon. There have been more than 40 attempts to study and visit our frequently overlooked sister Venus, 40 missions to Mars, 2 to Mercury, and nearly 40 to observe and monitor the Sun—often from the safety of Earth's orbit.

We've sent probes to Jupiter and Saturn, flown by Uranus and Neptune, visited asteroids, blown a crater in a comet's nucleus, and gathered up interplanetary dust—microscopic grains born here and also those drifting in from interstellar space. A mission is now on its way to Pluto and other trans-Neptunian bodies in the more distant reaches of our system. And the Pioneer and Voyager probes are even en route to the stars themselves, only now reaching the real start of their interstellar journey after a forty-year traversal of our local space, before settling in for tens of thousands of years of lonely passage.

Our own world has also been under constant scientific surveillance from space for the past fifty years, and we've succeeded in littering the vacuum near Earth with functioning satellites and a cloud of artificial debris. As I write this, there are some three thousand satellites orbiting the Earth, along with tens of thousands of pieces of junk larger than about a third of an inch, and tens of millions of smaller particles.

To some extent this exploration and occupation of space has always been driven by the hunt for other life. It's always been at the back of someone's mind, whether we've been examining the dense Venusian atmosphere, watching a Martian dust storm subside, or peering at the icy ridges on the surface of the Jovian moon Europa. Even the cryogenic lakes of methane and the eerily familiar hydrocarbon hills and valleys of distant Titan have prompted serious contemplation of the possibility of alien forms of life inhabiting these low-temperature environments, as I've discussed in the last chapter. But at the very dawn of our physical exploration of the solar system, back in the late 1950s, we didn't have a coherent idea of what to look for, and to a certain degree we still don't.

What has changed in the past decades is the growth of a more explicit acknowledgment that the search for life elsewhere plays a big role in motivating our exploration efforts. Indeed, it is often now the stated primary goal invoked to make a scientific case for funding and supporting new planetary missions. This focus has helped us refine our methods of exploration. As hunters of the big and the exceedingly small, we've become sophisticated toolmakers, with devices to seek out rare molecules and cameras to chart entire worlds.

Clearly, we don't know exactly what we should be sifting the sands of Mars *for*, or exactly *what* we are trying to spot lurking under the surface of Europa or Enceladus. When it comes to the ground truth for biology, we're very, very dependent on what we already know about organisms here on Earth, and this influences both what we think of as "life" and the ways we look for it. In the previous chapter I talked a little about the great "tree" of life on Earth, a branching classification of organisms that points toward distinct domains—the bacteria, archaea, eukarya, and possibly the viruses. The general consensus is that all of these domains share a common ancestry. Indeed, we typically discuss a "last universal common ancestor," or LUCA, a single species (conceivably even a single root organism, a grandmother to end all grandmothers) from which all subsequent life diverged billions of years ago.

Sophisticated statistical analyses (yes, Bayesian analyses) have been made of various scenarios for the ancestry of key pieces of genetic material common to all organisms. The results overwhelmingly support the notion of a LUCA, a single species that managed to evolve into all we know of life today, rather than a more complex ancestral arrangement. Exactly how this über-grandparent later gave rise to three or more distinct domains is not clear, though. The consensus is that bacteria and archaea came along before eukarya. That's a sensible proposition because, as we've talked about before, the larger eukaryotic cells include the subsumed parts of earlier simple-celled organisms. These absorbed symbiotes became organelles, such as the mitochondrion—a structure that is central to eukaryotic metabolism, and which I'll discuss later.

A great deal of careful study has been made on the possible specifics of a LUCA, from the requirements of its genetic molecular toolkit to its actual physical mechanics. But this is a messy business. For example, scientists who study the genetic divergence of life are still not sure that if we push the biomolecular clock further and further back, all branches in the tree of life really will neatly converge into one cleanly distinct species. Rather, there may be all sorts of incestuous back-and-forth in a small genetic pool that's still consistent with the statistical conclusions. Genes would transfer "horizontally" among individuals and nascent lineages in such a pool, and histories would become entwined and wrapped together in symbiosis or parasitic union.

Regardless of the details, eventually we also come up against what is broadly accepted as a likely transition from an earlier, pre-speciation form of life. That's a stage before a LUCA, which we think already had to have been a bona fide recognizable cellular species, DNA and all. Attempts to imagine this pre-LUCA stage are typified by the idea of an "RNA world," first put forth by Carl Woese in the 1960s. RNA is the "other" critical molecular structure in today's life, alongside DNA and proteins. In many respects it's a lot like a single-strand, shorter version of DNA, with some compositional differences. In other respects it's very different. It plays the central role of information transfer between DNA and proteins: strands of RNA are transcribed from the DNA code and can then be "read" by molecular machines called ribosomes that act like sewing machines to stitch new proteins together from the RNA information.

The postulated RNA world would be a sort of prototyping factory for DNA-based life—a smorgasbord of interacting structures at the dawn of cellular organisms. This kind of complex molecular ecosystem could represent a time much closer to the origin point of life, but it, too, must have been a product of evolution from something else. Perhaps that "something else" started with the first fatty lipids and cell membranes, and the first self-replicating molecules made from raw amino acid ingredients. We just don't know yet.

So as we push back to life's origins, the picture gets complicated in a hurry. We don't have real fossils of any of this early life from 3.5 to 4 billion years ago (although a team of geologists claims to have found fossil cells in 3.4-billion-year-old Australian rocks). All we have is chemical deposits and mineral structures left behind by colonies of single-celled organisms, or what came before them. As a result, we have to try to extrapolate from the molecular equivalent of fossils—items such as the protein structures encoded in modern DNA, each one like a microscopic layer of strata copied across untold quadrillions of organisms during the history of life on Earth.

This presents a knotty problem for answering questions about how many independent strands of life could have arisen on Earth, and how many abiogenesis events could have happened here or elsewhere in the solar system. The genetic fossil record does not have a perfect chronometer to match changes to an external time scale, and clearly we're also a little unsure of what would constitute a proper scientific definition of life's beginning. Long before a LUCA arises, we have to ask at what point we would consider a complex molecular structure "alive." This is a question that is as old as science, and we still don't have a very succinct answer to it, because life has multiple characteristics, from metabolism to reproduction and inheritance, and from homeostasis (the regulation of an internal environment) to adaptability. But there are some clues lurking in the biological undergrowth.

The peculiar case of the giant viruses offers one such insight. Viruses have long been considered "not quite life"—simpler packets of DNA or RNA that are utterly dependent on host organisms to supply the molecular toolkit for reproduction. However, nature may not be quite so easily classified. In the early 1990s, researchers studying amoebas from water in cooling and air-supply systems came across an organism infecting these small creatures. Initially thought to be a type of bacterium, it wasn't correctly identified until the early 2000s, when, examining it under an electron microscope, its discoverers realized that it was a virus—and a very large virus at that.

The original "Mimivirus" is about 750 nanometers, or .00003 (three hundred-thousandths) of an inch, in diameter, which makes it a giant among viruses. Not only is it much larger than most commonly identified viruses; this remarkable structure carries inside itself a significant set of DNA. In fact, it contains almost 1.2 million nucleic acid "letters," describing genes for more than 900 types of protein molecule. That may not sound like much, as human DNA contains up to about 25,000 protein-encoding genes, but in comparison the minimum genetic code that we've seen for a conventional virus consists of a mere 4 genes. Even some bacteria don't have this much genetic information in their makeup. Mimivirus is a beast. Since the discovery of these first giant viruses, more species (if we can use that term) have come to light, including one known rather affectionately as "Megavirus" that carries enough DNA to encode about 140 more genes than Mimivirus. This suggests that far from being an anomaly, the giant viruses are yet another template in the panoply of life.

Are they life? Do they deserve a new domain in the tree of life? Scientists studying the intricate details of the protein codes that the giant viruses carry have discovered some startling molecular evidence that helps answer these questions. Although the viruses still seem, like their smaller cousins, to depend on a host organism in order to reproduce and make use of their DNA, they carry genes for ancient protein structures that are also present in cellular life— bacteria, archaea, and eukarya. In addition, they contain enzymes involved in converting DNA code into proteins—enzymes that we had previously found only in cellular life.

This is not what we expect from viruses. It's as if these giant viruses are unemployed mechanics carrying around old toolkits. Although viruses can pick up genes from other organisms, for these giants to have obtained all these useful genes piecemeal seems unlikely. The remarkable conclusion is that these organisms may represent "de-evolved," or reduced, versions of something *else* that was once much more complex. They're almost, but not quite, capable of reproducing by themselves. But perhaps once upon a time they were. Somewhere along the evolutionary path they found a better existence as infectious parasites, or simply failed as something more self-sufficient. Some of the scientists researching these extraordinary viruses suggest that they might have originated from a different branch of life, predating or coexisting with the LUCA at the base of the other branches.

Only time will tell where this research takes us, but it raises some fascinating issues for our Bayesian interpretation of life's probability. Might we count something like the giant virus ancestor as a truly independent version of life? It appears to use all the same biochemistry as the rest of us, and could have arisen from the same earlier swamp of RNA or precursor chemistry. If it didn't emerge at precisely the same time as our LUCA, but rather a few tens or even hundreds of millions of years before or after, could we count it as an independent origin event?

Its de-evolved state may be telling us something else, too. It could be evidence that once life gets going on a planet, there is relatively little time for distinct domains with distinct biomolecular strategies to establish themselves before being outcompeted for energy and raw materials. If that's the case, it suggests that life on a planet is a first-come, first-served proposition. This means that it's unlikely that there are ongoing natural experiments with "new" types of life. They simply don't stand a chance in the competition for resources and a viable niche.

This leads us to an important question. Is the biochemistry of life as we know it unique *on Earth*? Perhaps there might be a way for a truly independent type of life, with an independent origin, to coexist with us today *if* it utilized a very different biochemistry. In other words, if it were able to avoid direct competition with all known organisms, it could be hiding in plain sight.

Some scientists, most notably the physicist Paul Davies, have taken a careful look at the notion of how such life might have managed to either completely evade direct detection, or somehow exist unrecognized among everything else. "Shadow life" could operate with a different chemical rulebook, one that drastically limits its physical and chemical visibility to us. That's the proposition, and it's quite a stretch, because the chemistry that known life operates with on Earth is so good at what it does. Finding an alternative molecular language that nature can build organisms with represents an immense challenge to our imaginations, and possibly to nature itself.

Any direct search for this kind of shadow life is obviously going to be tough. Since we're not aware of anything walking, slithering, flying, or swimming around that's built on entirely different biochemical principles, the more natural place to look is in the microcosm. But it's not that simple. Even now, a major fraction of what we know about normal microbial life comes from studying the genetics of populations, not individuals, and often not even individual species, but rather the genetic soup of many. Sieving through the contents of a pond, or the goop underneath a rock, is a painstaking business in the best of situations. If you're hunting for shadow organisms, not knowing whether your biochemical tests and analyses will even work, progress is going to be slow.

One trick might be to look for strange organisms surviving in conditions that should kill off "known" forms of life. We can let a toxic environment sift out the peculiar ones for us. The catch is that regular old life excels at adapting and surviving by the skin of its metaphorical teeth, a property that created a minor incident of media and scientific controversy in late 2010.

The story begins in an environment that is pretty weird even by the standards of the weirdest places we can find on Earth. It's a place called Mono Lake, situated on the eastern outskirts of Yosemite National Park in California, close to the Nevada border. Mono is a landlocked body of water that filled some 760,000 years ago. The lake's closed-off situation has combined with the local mineralogical and volcanic environment to produce water that's full of salts, and highly alkaline. This has been exacerbated by human activity: in the 1940s, water from many of the streams feeding into the lake was diverted to help quench the thirst of a growing Los Angeles.

With far less incoming freshwater the lake evaporated rapidly, becoming shallower and extremely saline—more than twice as salty as typical ocean water. Despite this, the lake harbors a remarkably productive ecosystem of brine shrimp, water-reliant alkali flies, microbial life, and colonies of birds feeding off these smaller inhabitants. It's an impressively diverse bloom of living things that obscures some of the highly toxic qualities of the water. For example, there's a high concentration of arsenic in the mountain runoffs still feeding the lake, which presents an enormous challenge to normal biochemistry. Arsenic is one of the most insidious elements known—if we were to assign it a behavioral characteristic. The problem with arsenic atoms is that they have a chemical flavor similar to that of phosphorus atoms, and phosphorus is a vital element in biochemistry. The arsenic atom is significantly bigger, but its outermost arrangement of chemically important electrons is the same as in phosphorus. As a result, if we ingest arsenic in the form of molecules of arsenate (arsenic and oxygen), these molecules can temporarily trick our system into thinking they are phosphate, with disastrous results.

Our bodies will mistakenly try to use this arsenate by incorporating it into various critical places, from energy transport molecules to even the backbone of DNA, where phosphate normally plays a crucial role. But while arsenate may smell like phosphate to our eager biochemistry, it doesn't function in the same way, and eventually these alien molecules disrupt and destroy cellular functions, killing the host. The chemical similarity of arsenic to phosphorus is still intriguing, though, and has led some scientists to speculate that instead of dying, some organisms might have evolved to operate with arsenic instead of phosphorous. Arsenic-based life could have arisen to occupy special ecosystems like those in the sticky mud at the bottom of Mono Lake. On paper, at least, this hypothesis sounds like a plausible candidate for shadow life.

But there are serious problems with some of the fundamental assumptions in this kind of idea. The extraordinary machinery of organic chemistry that constitutes "normal" life on Earth hinges on the precise physics of specific atoms and molecules. Replacing an atom with one of different size and mass will fundamentally alter the energy of bonds between atoms and molecules and the energetics of chemical reactions. Based on physics alone, it seems improbable that arsenic could replace phosphorus without some considerable rewriting of the biomolecular code for life.

Nothing beats going and looking, though, and in late 2010 a team of NASA-funded scientists published the results of a detailed study of microbial organisms in the arsenic-rich muddy deposits from Mono Lake. They had designed an experiment to coax out any organisms that were resisting the toxic effects of arsenic or actually incorporating it into biochemistry. To do this, they incubated the samples of bacteria and archaea in solutions containing ever-decreasing amounts of phosphorus and a high arsenic content. Remarkably, one species of bacterium, part of the family called *Halomonadaceae*—a salt-loving brigade—appeared to carry on quite well even when there was almost no phosphorus present. The scientists wondered whether this microbe might be doing something radically different, whether it was revealing a shadowy side. Could it be arsenic-based life?

What followed was a case of unwarranted scientific hubris, media hype claiming that this discovery changed the rulebook for life on Earth and beyond, and a colossal amount of speculation and gossip. Just before the news broke I was fortunate enough to be able to read over the forthcoming press release from NASA. To a quick glance it looked tremendously interesting. The scientists seemed to be saying that they had firm evidence that this species of bacterium was not only resisting the toxic effects of arsenic, but was incorporating arsenic into its DNA and still working properly. My quoted reaction to the scientists' claim was, "It's like if you or I morphed into fully functioning cyborgs after being thrown into a room of electronic scrap with nothing to eat."

However, almost immediately after the scientific report was published, microbiologists began to see problems with some of the analyses. Some of the claims being made to the media also were not well substantiated in the report. But it was far from simple to replicate and examine the results: this was a previously unknown species, and a substantial number of tests and experiments would be needed to reproduce the study. It was not a golden moment for science. Personalities got in the way of progress, and some journalists fed the fires of hype and controversy in their quest for the definitive story.

Eventually the furor passed, and scientists from other laboratories were able to take an independent look at the experiment. I think it's fair to say at this point in time that the overwhelming scientific consensus is that this brand of bacterium is incredibly resilient to arsenic toxicity, but it's not arsenic-based life. It's very good at finding ways to survive despite being bathed in the stuff. It may even manage to incorporate arsenic as a functional replacement for phosphorus in a few processes—but these end up being far less efficient than the usual phosphorus-based ones. And there is no evidence that arsenic operates like phosphorus within the very DNA of the bacteria. Indeed, take every last crumb of phosphorus away, and just like all other known life, these bacteria will die.

Specifically, a study from 2012 found that the proteins in this bacterium that are responsible for pulling in phosphorus-containing molecules from its surroundings have a four-thousand-fold preference for these molecules over equivalent ones that incorporate arsenic. In other words, and rather ironically, this organism is actually very, very good at picking out phosphorus when it finds itself engulfed by a sea of arsenic. Such fussy habits help the bacterium stay alive while others succumb to the destructive effects of arsenic.

It's too bad. It would have been an amazing discovery to find an example of shadow life, but it doesn't look like this is it. Rather, it is a cautionary tale that highlights the challenges of digging around for shadow life under our noses, life that might operate differently and descend from a genuinely independent origin. And the difficulty of doing this may itself be telling us something important. Why is it so hard to look, and so easy to be fooled?

That question brings us all the way back to examining the ways in which we infer properties about the universe around us, including our own status. Thomas Bayes's theorem tells us that right now we're lacking critical information, and that includes whether or not life has started independently more than once here on Earth or anywhere else in the cosmos. We have a lot of evidence that known life fits very well into the elemental and chemical composition of the cosmos, and we've found that the cosmos produces an abundance of planets. But we've yet to succeed in linking the fact of our existence to all of this in a quantitative way. I would argue that we make more progress by extrapolating downward from our knowledge of the rich universes of interstellar molecules and planet-forming processes. It's easy to see that life's properties on Earth are connected to that set of cosmic conditions. Going in the other direction—extrapolating upward from what we know and surmise about life's origin on Earth, and trying from that to predict the likelihood of life elsewhere— doesn't seem to yield much insight. Our past efforts along these lines have led to conjectures that are polar opposites of each other, from anthropic uniqueness to the plurality of worlds. Modern Bayesian inferences about cosmic abiogenesis lead us back to square one.

Of course, it's inevitable that our inferences about the cosmic existence of life are somewhat based on our own circumstances, but it's also extremely hazardous to work this way. To avoid this trap, we need to be aware that our perspective on the cosmos is itself a product of our status and circumstances. Perhaps the blinkers on our restless eyes are larger than we've realized, and we need to try to take them off.

7

THERE'S SOMETHING ABOUT HERE

Imagine, if you will, that the Earth had formed around not a single star, but twin suns. This isn't just the stuff of fun science fiction anymore. We know that planetary systems like this exist—pairs of closely orbiting stars with planets orbiting outside of them. In one place, called Kepler-47 after the NASA observatory that discovered it, the pair of stars complete an orbital cycle every seven and a half Earth days. Safely outside their stellar waltz are at least two worlds following slower, statelier orbital paths.

It is of course impossible to accurately predict how human beings in such circumstances would deal with what they witness in their skies. But with a little imagination a number of possibilities come to mind. (For convenience, let's assume that this alternative Earth rotates more or less as ours does.) First, as they watched their twin suns cross the daytime sky over a week or so, they would see these two brilliant disks passing back and forth relative to each other. If the geometry is perfectly aligned, then the two stars would take turns eclipsing each other at certain times. This other Earth therefore experiences nights, days, and two types of eclipsed days that come periodically, approximately twice a week.

How would we assemble our cosmological thoughts if we lived in such a system? Well, there would be a number of significant factors to consider. For example, as this alternate Earth orbited its twin suns, the time at which the star-on-star eclipses take place would shift. This drift would happen with the passage of a year, and if the planet's axis were tilted, as our Earth's is, the timing of the solar eclipses would noticeably change in relation to every solstice. It would be a tricky variation, begging for a clear explanation.

As alien as this system is, though, a geocentric picture (in which this planet is thought to be the center of the cosmos) could still be made to work. The stars could simply move around each other much like they would on an epicycle in Ptolemy's cosmology, and the center of that circle would move around this Earth in another great circle, or deferent.

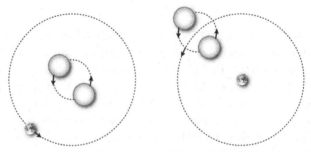

Figure 13: The twin suns of an alternate Earth. Even though the true configuration is centered on the stars (*left*), an intelligent species could still construct a model that would accurately describe what they saw in their skies, and let them believe that their planet was central (*right*).

With some geometric tweaks to this geocentric model, the prediction of when these suns eclipse could be adjusted to drift by the right amount to stay in step with the seasons. Just as on our real Earth, the biggest clue to a proper heliocentric arrangement (remember Aristarchus) might come from the movements of any other planets in the system, looping back and forth across the sky.

Even this twin-sun system might not give its inhabitants any clearer hints about their significance than ours does. They, too, would have to wait for a Copernicus to set the record straight and decentralize their home. But this is just one example. Consider another scenario.

We can imagine the possibility that this Earth is a small planet within a densely packed system of much larger planets, where rocky giants and gas-rich giants populate the inner orbital zones. Based on what we've seen of exoplanets, this sort of setup likely occurs more often than the configuration of our actual solar system. Let's suppose that this planet-rich system contains eight other worlds between our hypothetical Earth and the Sun. All of these are larger than the Earth, and some of them are the size of Neptune. Such a place is not pure conjecture. Some recently discovered exoplanetary systems have just such arrangements. We do not yet know whether they harbor anything precisely equivalent to our home world, but we do know that they could.

In this scenario, these interior planets appear as brilliant objects in the sky as they drift through it night after night, looping back and forth, appearing and disappearing as the weeks and months go by. The larger ones are sufficiently big so that even the naked eye can discern the changing crescent-shaped illumination by the Sun—no need for another Galileo to build a telescope to see this phenomenon.

Faced with all this variation, our hypothetical cousins wouldn't think the planetary movements are mere inconsistencies. Far from it. The inhabitants of this Earth would soon figure out that all the action is centered around the Sun. This wouldn't diminish their sense of self-importance in the slightest. They started out convinced that the Earth is significant—after all, it is clearly placed in the best seat in the house to watch the great clockwork motions of the inner, *inferior* worlds. Their civilization would still have scant evidence for precisely how distant the tiny, unwandering stars are that they see at night. But these luminous pinpricks never resolve into planet-like disks, so they must be a long way away if they are other worlds. And if instead they are other suns, it is also obvious that their planetary systems cannot be seen only because of their great distance. Atomists and pluralists would prevail in the natural sciences on this world, with their theories positing a wealth of other planet-laden systems in the universe. After all, some truths really are self-evident.

There is another possibility that would turn our own history of cosmic discovery on its head. What if we imagine an Earth that was never a bona fide planet at all, but instead was a place further down in the hierarchy of worlds? What if our home world was actually a home *moon*? It's entirely plausible for a gas giant planet to harbor "home moons"—big enough to hold on to an atmosphere, big enough to behave like planets themselves. Titan in our own solar system easily meets this criterion, and larger moons could exist elsewhere. If the host giant orbits a Sun-like star at the same distance at which we orbit our Sun, such a moon would be heated similarly to the Earth, and could conceivably harbor a similar surface environment. It's a complex scenario, but one long favored by science-fiction writers and filmmakers. For the development of a civilization, it's an intriguing hypothetical case.

The most likely physical configuration for a moon around a giant planet is to be in a state of synchronicity between its rotation and its orbit. In other words, the moon will always face its mother planet the same way, the time it takes to make one rotation matching the time it takes to make one revolution around the planet. Our own Moon is in this situation, and the cause has been the relentless drag of gravitational tides across the eons. These small pulls and tugs can gradually coerce moons to slow their rotation by bleeding away whatever original spin they had until it matches their orbital period.

So one entire hemisphere of our imagined home moon will always face the giant planet, a huge looming thing covering perhaps 20 degrees of the sky, roughly equivalent to a pair of hands held up at arm's length. The other side of the home moon, the far side, will never see the planet behind it, always facing out to open space. Indeed, the first modern explorers from that opposite hemisphere will be startled to see the ominous planetary orb rise above the horizon as they trek across the surface.

The passage of time is marked by a number of extraordinary displays on this moon's near side, facing the gas giant. During a semi-dark period of night the mother planet hangs brilliantly in the sky, flooding this satellite world with reflected sunlight—planet-shine. And, in the scenario I'm imagining, a near-perfect geometrical alignment of orbits and axial tilts results in the shadow of this moon-

Earth falling directly onto the great planetary disk. Evident to even the casual observer, this dark spot tracks slowly across the gas giant.

It's a crucial marker for the sentient inhabitants, because when this shadow reaches the edge of the planet's disk, it portends the end of a planet-shine night and the start of a new day. Gradually the sun begins to crawl up from the horizon. And as if there were some magical connection across the void, the disk of the great mother planet at the same time begins to turn into a narrowing crescent, the terminator of growing shadow curving across what is clearly the surface of a giant sphere.

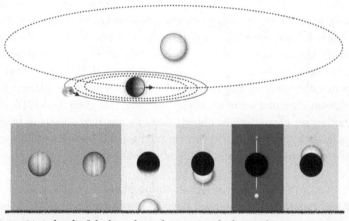

Figure 14: A sketch of the hypothetical system in which "Earth" is a moon around a giant planet. *Top*, the orbital configuration (not to scale) showing the giant planet orbiting its star and the paths of some of its moons, including an alternate Earth. *Below*, a schematic of some of the phases of the giant planet and star as seen by an inhabitant on the near side of the moon-Earth. *Left to right*, we pass through planet-shine (sunlit giant planet hanging stationary in the sky) to full day (star rises above horizon), to full night (star is eclipsed as it appears to move behind the fixed giant planet). Although not to scale, a pair of other moons (out of dozens) and a thin ring system become visible as we pass through the different phases.

Generation upon generation of adept geometers and mathematicians have been produced on our hypothetical home moon, motivated by observations of this shadowing. But nature is not finished yet. Now, after a brief full illumination, the near side of the moon is about to slip into a far deeper night. This second, alternating night begins as the disk of the planet in the sky—still immobile in its place—starts to blacken to nothing but a luminous crescent. The sun, having moved across the sky toward the great planetary disk, now dips behind it and is eclipsed.

Darkness falls across the world, and in the blackening sky the distant stars appear even brighter, except where they are blotted out by the disk of the planet and its ghostly halo—a faint glimmering ring of sunlight—where the planetary atmosphere refracts and reflects the sun's glow. Poetry has been written of this ghostly halo, and still is, even after science demonstrates its origins. It is in this second night that another phenomenon is also at its most vivid. Always there, but previously outshone by planet and sun, a thin line of light—we can now see—cuts through the sky. It juts out from either side of the blackened disk of the mother planet: the edge-on view of a set of icy rings encircling this giant world. There's another set of mysterious objects that emerge, too: dozens of bright spots—a few discernible as tiny round disks. These beads of light are splayed across the sky in alignment with the thin line of the planetary rings.

Centuries earlier, the great philosophers and astronomers of this alternate Earth proposed that the beads were nothing less than other moon worlds like our own. In fact, as they measured their motions and brightness, they determined long ago by geometry and logic that the mother planet was indeed the immediate central celestial hub. Astute astronomers also found that there were specific relationships between the time it would take for these objects to complete a cycle around the mother planet and their distance from it. Not only that, but they realized that their own world's motion around the mother planet followed the same rule.

In this hypothetical place it took only a few steps of reasoning to establish a universal law of forces that depended on the quantity called mass—a law that came to be known as gravity. As bigger and better telescopes became available and scientists tracked other distant and previously unnoticed planets (some with their own moon worlds), inhabitants of this other Earth quickly deduced that everything moved around the sun. This motion was readily explained by that universal truth of gravity. The occupants of the home moon revere their worldview for its beautiful hierarchical elegance. The sun is the grandmother, the mother planets move around the sun, and the daughter moons move around them—all governed by the same set of immutable physical laws.

It's clear that the beings of this imagined moon world have had it easy compared to our millennia of struggles over the cosmic order of planets and our status. Why is that? It's because when it comes to making sense of the universe, circumstances are everything. These circumstances are also profoundly connected to the chances of life existing in an environment in the first place—a great chicken-and-egg puzzle. Untangling this puzzle is the next step toward resolving the tension between the evidence for and the evidence against our cosmic uniqueness and significance.

All of these hypothetical worlds are (as far as we know) just thought experiments for illustration. Now let's come back to the real Earth. One thing that's surprising about the history of science and of our investigation of the universe is how often big insights have hinged on the most tiresome scientific minutiae imaginable. This in itself may be an important clue to our place in the cosmos.

Many of our discoveries about nature have come to light only after examining tiny details—the little nagging inconsistencies that at first seem so technical and esoteric and only later seem so marvelous. Breakthroughs have happened when someone became intrigued with the slightly wayward motions of planets, the oddly constant speed of light, or the subtle variation of living species and the layers upon layers of confusing fossils. It takes a robust constitution to stomach working on these kinds of problems, and it's mostly fallen to a unique cadre of restless tinkerers and worriers across the ages. Sometimes they've taken great personal delight in their painstaking tasks, to the despair and annoyance of their colleagues. Sometimes it's taken a while for the general population to even understand what the fuss is all about. The Copernican revolution in thinking is an excellent example of how excruciatingly dull details can also make for the most dramatic and upsetting revolutions. Nicolaus Copernicus's last and most important work, the great *De revolutionibus*, is so packed with astronomical technicalities that only the best-trained astronomers of the time would have found much to be excited by. Indeed, its hairy complexity surely helped spare it from worse criticism by church and state. As the twentieth-century physicist and philosopher Thomas Kuhn rather dryly put it in his history of the subject, "Opposition to a more comprehensible work might . . . have been marshaled sooner."

And it's not too surprising, because to a large degree Copernicus's motivation was to improve the basis of existing models of celestial mechanics, not just to make an elegant argument. In this respect he was akin to an obsessive trainspotter, only he was trying to generate a more accurate timetable of the positions of the planets in the night sky. Resetting the structure of cosmological thought may have been a by-product of these efforts (although he clearly understood the implications). Sorry, Nicolaus, we love your work, even if it isn't exactly bedtime reading.

Over half a century later, the mathematically possessed Johannes Kepler was similarly driven, working for no less than eight years on the orbits of Mars and the other planets. As determined as he was to pin down the "clockworks" of planetary motions, and to discern what might be responsible for them, he also just wanted to get rid of the annoying inconsistencies, the varying planetary brightnesses and slightly mistimed positions, that plagued earlier astronomical systems.

And even when Galileo finally saw the movement of moons around Jupiter, the innumerable stars composing the haze of the Milky Way, the shadowed landscape of the Moon, and the crescent-shaped illumination of Venus, these were still all pieces of the puzzle—small clues to an emerging worldview. The genius of all these people was in what they were able to extrapolate from such details: a heliocentric cosmos, the true shape of orbits, and the nature of motion and forces.

So we can see how our evolving understanding of the universe and our place in it has depended on particular circumstances of the planet Earth, our solar system, and its position in space and time. Of course it's easy to look at the history of science and think that we know better now, that we no longer have such tunnel vision. We might assume that the precision of observation and measurement available to us with modern technology has lifted us above the mire of all of these earlier struggles with nature's details. We can measure celestial positions to thousandths of a degree now, or gauge distances and velocities across billions of light-years. But the truth is that we are still hemmed in and blinkered by our own circumstances, for studying both the cosmic and the microscopic.

Back in the first chapter I asked what would have happened if the history of astronomy had played out differently—if Galileo had gone on to build enormous telescopes and discover life on other worlds. That was pure fantasy, but now we know that our galaxy, and the universe as a whole, is swarming with other planets. We also know that the diversity of these worlds and the range of their configurations and histories lends statistical weight to the idea that our planetary circumstances are unusual. And, as I've tried to illustrate with these fantasies of human life on other worlds, this means that our outlook might be unusual, too.

The first question that springs to mind is whether our perspective has helped or hindered the development of our scientific method itself: What could the blind spots be hiding from us right now? The second question is more unsettling: What if the very configuration and history of our planetary system that has made life possible on Earth has *also* placed severe constraints on the way we've developed a picture of the universe? To put this another way: Is life like us always going to ask the same kinds of questions because it can only exist in the same cosmic circumstances?

The celestial scenarios we've imagined here, from twins suns to Earth-like moons, are plausible in terms of physics and astronomy.

What we don't know is whether or not they're plausible in terms of biology. For one thing, we don't know whether these speculative planetary environments would have characteristics that might make it hard for life to arise and evolve. For another, we have no theory that could predict what kind of sentience might evolve, or how the unsteady passage of chance and history would affect its interpretation of the surrounding cosmos.

There is no doubt, though, that for *us* an alternate set of planetary circumstances could easily have led to a different development of natural philosophy, and a radically alternative scientific history. For better or worse, our own worldview has at times been trapped in a rut because some of the most important clues are buried in the details of what we see around us. But this might be the case in any planetary system capable of sustaining life over the long term.

For a thought-provoking example, we can come back to Johannes Kepler's analysis of the orbit of Mars. You'll recall that he made an inspired choice by studying Mars's wanderings in the sky because it has the least circular orbit of the major planets, except for Mercury. But his choice was also inadvertently made for him by the vagaries of time and celestial physics, because we've since learned that Mars's orbit wasn't always, and won't always be, what it is now.

In fact, just as the overall orbital dynamics of our solar system dance very close to chaos, Mars's orbit shifts over time—influenced by the gravitational tugs of the other worlds, especially Jupiter and Saturn. The ellipticity of the Martian orbit can change significantly, oscillating over a period of about 96,000 years by as much as a factor of two. Over even longer timescales, millions to tens of millions of years, it can veer from almost circular to nearly double the ellipticity it has today.

In other words, had humans come along a hundred thousand years earlier or later, if there were still a Kepler studying a Brahe's charts of planetary motions, he might have had either a much harder task or a much easier one. If Mars had been in a nearly circular orbit when Brahe made his measurements, it wouldn't have offered *any* clue to the more general nature of planetary motions. Equally, if it had been in an even more elliptical orbit, someone else might have beaten Kepler to the prize.

But this kind of orbital behavior—changing shape and angle and other parameters over time—is tightly correlated with the overall architecture of our solar system and its deeper history, as we've discussed. Earth's orbit and rotational orientation experience small, slow changes, too. These combined shifts in configuration appear to be correlated with longer-term changes in terrestrial climate, including deep ice ages that occur on 100,000-year cycles. It's an intriguing possibility that at many points in history when orbital conditions for Mars would have favored easy measurement of its ellipticity, the temperature conditions on Earth would have been unfavorable for a species like humans.

There are other variations in the physical environment on a habitable planet that could radically alter our perception of the universe. If we had an atmosphere clogged with condensing water, or with the haze caused by the photochemistry of organic molecules such as methane, we might never make precision measurements of anything other than the Sun or Moon. And it's quite possible that there have been periods on Earth when millennia of nothing more than bad weather would have hampered our efforts to study the heavens, if we'd been around.

The galactic environment of a solar system can stifle its view of the cosmos even more. We know that the Sun and its planets follow an orbital path around the Milky Way galaxy that completes a circuit every 230 million years or so. But this path is not a perfect circle or an ellipse, since the galaxy itself is a landscape of undulating concentrations of mass and complex gravitational fields. Furthermore, none of the components of the galaxy are stationary; they, too, are orbiting and drifting in a three-dimensional ballet.

The result is that our solar system, like billions of others, must inevitably encounter patches of interstellar space containing the thicker molecular gases and microscopic dust grains of nebulae. It takes tens of thousands to hundreds of thousands of years to pass through one of these regions. This may happen only once every few hundred million years, but if modern human civilization had kicked off during such an episode, we would have barely seen more than the nearest stars—certainly not the rest of our galaxy or the cosmos beyond.

But could our planetary circumstances have been that different and still produced us? Would more changeable orbits in a planetary system, or bad weather, or passage through interstellar clouds, also thwart the emergence of life in some way? Phenomena such as these could be bad news, causing hostile surface environments on a planet. So it's a possibility that the planetary requirements for forming sentient life like us will necessarily always present the senses and minds of such creatures with a specific cosmic tableau, a common window onto the universe. If that sounds familiar, it's because the premise of the anthropic principle is the same: that an observer will see only these types of surroundings because they are necessary for an observer to exist in the first place. In this case the basis for the idea is much more parochial, and the solution may be more straightforward.

The relationship of life to its planetary environment brings us back to the perennial, gnawing question: Just how rare or common is life like that on Earth? Biologists often split this puzzle into two pieces, one concerning "simple" life and the other "complex" life. Other scientists tend to conflate both in the word "life." But the strict distinction between "simple" and "complex" is one that we've encountered before in comparing bacteria and archaea with eukarya—the three great domains of life on Earth. Eukarya, such as us, have "complex" cells that are larger than those of bacteria or archaea and contain more, and more intricate, structures. Critically, they also keep their DNA wrapped up and protected inside little membrane sacs—cell nuclei. We think this complex cellular form arose later than that of the "simple" bacteria and archaea, and without it there would be no organisms like us walking around on the Earth.

The fact that there are simple and complex organisms feeds an idea that I noted back in the prologue: the "rare Earth" hypothesis. It could equally be called the "rare complex life" hypothesis, because it hinges on the notion that complex-celled life may be extraordinarily unusual in the universe. By extension, thinking and technological beings would also be few and far between. The proposal that complex-celled life is rare is an important idea to explore, but I'll remind you of a comment I made before: I don't think the basic premise of a rare Earth is well substantiated.

The idea of a rare Earth is that specific chains of planetary events and properties must all line up for complex-celled and perhaps intelligent life to evolve. Simple life (such as rock-eating microbes), on the other hand, might come along far more readily. To make this argument, we can draw on a wealth of detail about the Earth's history and circumstances. For example, take water. This simple molecule of two hydrogen atoms and one oxygen atom is a vital biochemical solvent and a central component in the machinery of geophysics on Earth. But the amount of water on the planet, and whether lots of it is in a liquid state on the surface, where complex life can utilize and enjoy it, is affected by many specific events and situations.

We could argue that water's presence on Earth is connected to the configuration of asteroids, comets, and giant planets in the solar system, as well as to the past and present evolution of Earth's orbit. In addition, complex life likely benefits from the presence of a protective planetary magnetic field, which is in turn related to the way the Earth-Moon system formed, and may even be maintained by the tidal pulls of the Moon. Without a relatively large moon, the Earth would also experience far greater variations in its axial tilt, and with them, violent changes in climate that could challenge complex life more than hardy microbial life. And the evolving composition of Earth's atmosphere and oceanic chemistry across time also undoubtedly relates to various quirks of geophysics, some of which date back to presolar, protoplanetary times—right down to the heating effects of radioisotopes from local supernovae that were folded into the coalescing mix. Indeed, without the action of terrestrial plate tectonics, which depends in part on Earth's internal heat, the chemical and topographic environment at the planet's surface, the interchange among continent, seafloor, and ocean, would be very different. And so on.

If we overlay these factors on the time line of all species' development over the past 4 billion years, life's evolving structure begins to look like a house of cards. Change any one thing here or there and it's possible to thwart the chain of events that has led to complex life, and to life like us (on a larger timescale, as hazardous a path as our immediate ancestors took when they emerged from Africa a hundred thousand years ago).

This is the crux of the rare Earth argument: The emergence of complex and intelligent life here on Earth was critically dependent on some, or even all, of the above-mentioned characteristics. Furthermore, there are few, if any, other workable options. If this is true, it means that only near-exact twins of Earth can harbor life at all like us. In other words, complex life must be unusual across the cosmos—even a cosmos filled with planets.

Even the apparent unlikelihood of our planetary conditions and history might not be the biggest factor for a rare Earth. Some scientists argue purely on the basis of biology that it's highly unlikely for complex organisms to arise anywhere at all, because of the specific chain of events necessary for vital bits of molecular machinery to wind up in the right places at the right time. The implication is again that complex life has to be very rare in the universe, and that special circumstances are necessary to help it emerge at all.

A central piece of this biological argument is that bacteria or archaea cannot easily "upgrade" to larger and more complex physical forms because they're simply not efficient enough at processing energy. The more genes an organism has, the more energy it requires to convert this genetic information into proteins. Stuck with their relatively basic methods for generating energy, microbial individuals can't afford to carry around a huge library of genetic material, and so they stay simple.

As I've discussed, eukaryotic cells are different from the hordes of single-celled organisms, and here's another reason why. They can contain extra structures called mitochondria—membrane-wrapped packages of DNA, RNA, and hundreds of enzymes. These packages are separate from the cell's own nucleus that protects the organism's primary DNA. Mitochondria are amazing things. Among other functions, they serve as specialized chemical power plants for eukaryotic life, performing oxidation to produce vital molecules that act as potent carriers to shuttle electrical energy around the cells. They are at the root of why we must breathe oxygen—and why we can grow to such great sizes.

Mitochondria make life like us possible because they turbocharge our metabolic efficiency. The energy they provide yields a

200,000-fold increase in the number of genes our cells can express compared with microbial life. But mitochondria are almost certainly bacterial in origin. We think that about 2 billion years ago they merged with the precursors of eukaryotic cells, becoming endosymbiotic—completely subsumed within their hosts, serving as critically important energy generators.

So far so good. But some scientists, such as the biochemists Nick Lane and Bill Martin, have argued that the organism that merged with the mitochondrial ancestor may not have been any more complex itself. They have asserted that the complex cells of eukaryotes *started* with the combination of two similar organisms. According to Lane and Martin, the entire history of complex life hinges on a single, random, and very unlikely cell merger.

To my mind, this may be the strongest argument yet made that the origin of complex-celled life on Earth was an extremely lucky event. It parallels, but supersedes, the astrophysical or planetary arguments that propose a connection between life and a chain of specific circumstances. If that concatenation of circumstances is required for life, that argument concludes, the odds are slender for there to be lots of planets enough like the Earth. But it lacks the apparent finality of the mitochondrial argument.

There are other striking cases of major endosymbiosis (one organism living permanently and peacefully inside another, to the benefit of both) on Earth. For example, chloroplasts—structures central to photosynthesis—in plant cells indicate that similar mergers have occurred. Scientists think that these bean-shaped microscopic structures originated as cyanobacteria, ancient photosynthesizing microbial life-forms. But plants, which also contain mitochondria, came along much more recently than complex cells. In fact, all evidence suggests that nothing quite like the "mitochondrial event" has happened again, not since that *one* time a couple of billion years ago.

The theory is compelling. But we don't actually know that the mitochondrial ancestor merged with just another simple-celled species. If a proto-eukaryote life-form already existed (perhaps even a type of more complex archaeon), the mitochondrial event would have been just a step in the evolution of that organism, another instance of an unremarkable type of endosymbiotic event that had happened before, when the proto-eukaryote engulfed helpful microbes and kept them around instead of digesting them. This would make the mitochondrial event far less pivotal, which I'll admit I prefer. I get twitchy with lines of reasoning that depend on "improbable" events. Some of these are suspiciously like the arguments of some twentieth-century scientists, such as the physicist Fred Hoyle, that the origin of life on Earth needed an external "seeding event." The proposal was that terrestrial biochemistry started with an organism that (while entirely natural, like a bacterium) came from beyond the Earth. We refer to this concept as panspermia—or "all seed" from its Greek root.

As Hoyle and others saw it, if you mixed atoms and molecules in a pond somewhere on a primitive Earth, the chances of an RNA or DNA molecule spontaneously forming within even a few billion years are almost nonexistent. Accordingly, life couldn't have started here by itself and must have been initiated by the arrival of some form of life or proto-life from elsewhere. The problem of abiogenesis was outsourced to the cosmos.

Today we think that the ways in which both basic and complicated molecular structures assemble, and the ways in which order can spontaneously emerge from complex systems, are much more efficient than we used to imagine. We also think there are a variety of nonliving, inorganic chemical and physical templates that could have helped push carbon chemistry toward becoming fully fledged biochemistry on a young Earth. Invoking panspermia as Hoyle envisioned it seems unnecessary. So while our modern ideas don't explain everything, they do suggest that we should be wary of assuming that an incompletely understood biological phenomenon is inherently unlikely to occur. So, in my opinion, to posit that complex cells with mitochondria have only a remote chance of popping out of the microbial stewpot is, at least superficially, to buy into the same old panspermia idea that life must be improbable because it seems so to us.

I'll go further. If the history of science teaches us anything, it is that we need to be more than wary of this kind of assumption. We can and should ignore the most extreme of such ideas. The reason is that they're often influenced by a misleading intuition about the nature of statistics. In fact, one of the most damning arguments against any version of a "rare Earth" comes from a relatively simple but powerful examination of the nature of probability and our perception of randomness.

There's an old story that statistics professors use to produce shock and awe in new students, and while it appears simple, it also highlights some entrenched habits we have about putting information into context. Like so many good stories, it gets told in a wide variety of ways. I'll use sports to illustrate the idea.

Joe is sitting at home one evening when he gets a phone call. To his surprise, the call is from a good friend whom he hasn't spoken with in more than five years. They talk, and the friend tells him that he happens to have a spare ticket to a big baseball game that night. Would Joe like to go?

An hour later Joe is making his way to an assigned seat in the stadium. Fifty thousand other fans are packing the seats. It's a full house, and everyone is shoulder to shoulder. When Joe and his friend get to their spot, someone asks if they'd mind swapping for slightly better places about fifteen feet away, so that a family group can sit together. Joe says no problem, and they take their new seats.

The game starts up, with a famous batter at the plate. As Joe gets settled in, he spots a snack vendor walking the stands and calls him over. Just as Joe is reaching for an item from the vendor's tray, the batter down on the field swings and hits it into the stands. The ball sails way up and toward the spectators, comes crashing into the vendor's tray, and bounces right into Joe's hand. It's a history-making hit, and the ball is instantly transformed into a piece of sports memorabilia.

Joe shakes his head: he can't believe how lucky he is. If his long-lost buddy hadn't called him, if he hadn't had a ticket for the big game, if they hadn't swapped seats, and if he hadn't been reaching for a snack from the vendor at that very instant, he wouldn't have caught the ball! It seems so incredible that for a while he wonders if he hasn't picked up some weird talent, becoming a human lightning rod for mysterious coincidences.

It seems reasonable. If something like this happened to one of us, it would definitely give us pause. We'd wonder whether the cosmos had picked us out especially for this event. After all, what are the odds of all those events lining up so perfectly?

The problem is that our intuition is very, very misleading when it comes to matters of randomness and chance. From Joe's perspective, this seems like an incredibly improbable chain of events. Out of fifty thousand people in a stadium, the ball got to him in just the right place at the right moment. But is Joe's perspective the most appropriate?

Not if you want to understand the global significance of the event. The point is, that powerfully hit baseball was going to whack into, or be caught by, *someone* in a packed stadium. It was inevitable. If it hadn't bounced into Joe's hand it would have bounced into someone else's, or knocked someone else in the head, or smashed their drink out of their hand. And each of those people would find themselves having the same kinds of thoughts as Joe. How amazing was *that*?

Each and every possible scenario would afford ample scope to gasp at the astronomically small odds of the ball landing where it did. In each and every possibility there would be just as much room to gasp at the extraordinary chain of events: the last-minute decision to go to the game, the impulse to look up at that instant, the lucky shirt that was worn, the hot dog that was halfway to the mouth. But these are all events that take on meaning only *after* the fact—they are what we generally call a posteriori information, or part of a post hoc analysis.

The true significance of what happened to Joe is not so simple to evaluate, but one thing is clear: it is not going to be as extraordinary as we might first think. Yes, it was extremely unlikely to happen to this particular Joe, but it wasn't at all unlikely to happen to *somebody* sitting in the stadium.

•

What does this have to do with the notion of Earth being a rarity as a harborer of complex and intelligent life? That idea is rooted in a posteriori knowledge. This is true whether we're considering the amazing steps of molecular biology involved in producing complex life, or the remarkable chain of astrophysical events leading to the modern Earth. You and I standing here marveling at the apparent near-miracle of our existence is not so very different from Joe marveling at the odds of his lucky catch.

We can dissect the various components of Earth's history and properties as much as we want, from the randomness of planet and moon formation and our geophysical and ecological history, to the steps and leaps of biological evolution—all the properties that appear to make our local patch just right for life. We may indeed find that each piece is important and cosmically rare, perhaps entirely random to boot. But this does not tell us for sure that our existence as complex, intelligent life-forms is itself a rarity in the universe.

In fact, it could be the polar opposite. Let's suppose that the rise of life, and the evolution of some of it toward increasing complexity and humanlike intelligence, is essentially inevitable wherever life can get a foothold. It's like the inevitability of well-hit baseballs hitting fans in a capacity crowd at a stadium. This doesn't mean that life can *always* manage to evolve to a complex level, but if the window of opportunity is large enough, it will.

From this perspective, our presence on Earth could have come about equally well by way of an immense number of alternative histories—and again, the particular characteristics we find in our one case are all a posteriori facts. Life like us (that is, complex-celled, dextrous, cerebral, linguistic, and technological) was going to happen given the slightest chance. How "lucky" we are is irrelevant. The participant in an inevitable event with a highly specific but random outcome will always think that he or she is a one-in-a-trillion miracle.

To put this another way, any thinking life-form anywhere in the universe may always perceive "special" characteristics in its own circumstances—specifics that, if different, would derail the chances of complex life occurring. This perceptual bias may be irresistible regardless of whether complex life is rare or as common as muck. Until we either discover life elsewhere or somehow rule it out, any post hoc interpretations of the significance of our circumstances are almost meaningless. A simplified way of looking at this is to ask the following: Does it make sense to claim that the existence of an object sitting in front of you is extremely improbable, or instead to say that your knowledge of how this object came to exist might be incomplete? I know which I favor.

We all feel strongly about the theories we adopt, but it's important to be clear that we're not talking about actually answering the questions of whether life is rare, or whether Earth is a rare haven for life. We need more information for that. Here, we're just saying that regardless of the answer, it will always *seem* that we've caught the proverbial one-in-a-trillion baseball.

I've already discussed how the nature of biological intelligence is not well enough understood to assume that it is a one-flavor-only rarity rooted in a chain of unusual events. Creatures like the fabulous cephalopods could yet prove this to be a fallacious assumption. The same difficulties arise deeper down the chain, with the planetary conditions needed for complex-celled life. How necessary were the specific celestial and biological circumstances we humans find ourselves in? I'm more comfortable assuming that they need not have been exactly like those on Earth. We'll have to wait to see if we can actually confirm or disprove this notion.

When we think about alternate planetary origins and scientific histories, and seemingly improbable biological events and evolutionary pathways, we're bumping up against the frontiers of knowledge. In the next and final chapter we'll see where this leaves our quest. Before that we'll make one more journey, one that offers a quick peek at the furthest reaches of our search for cosmic significance: the question of the very nature of reality itself, the nature of the universe as a whole and the place of conscious intelligence in it. Such epic subject matter also holds hidden pitfalls.

We can go in two directions to try to understand the nature of reality. One is inward, to the microscopic, the molecular, and, deeper still, the quantum world of matter and energy. The other is outward, to the grandest scales of space and time encompassing all the stars and galaxies, matter, dark matter, and cosmic radiation.

But although these directions are opposite, they are not separate—not at all. In fact, it is remarkable what the inner universe of matter and energy teaches us about the outer universe, and vice versa. The simple reason is that the most fundamental components of reality all boil forth from the same box of tricks. The physics of the grand architecture of the universe are the same as the physics of the molecular and subatomic realms. And we don't need to go very far into this incredible science to find a major clue to why our specific circumstances critically affect what we can and can't learn about the cosmos. It turns out that the universe is not always amenable to revealing itself to an observer.

In cosmology, one of the great discoveries of the past two decades is that the expansion of the universe—that swelling growth of the very fabric of space that propels galaxies apart from one another—is *accelerating*. Putting it in crude terms, instead of all the matter in the universe providing enough gravitational pull to eventually slow the expansion from its initial rush at the Big Bang, the universe is now increasing its rate of expansion. To confirm this behavior, astronomers have painstakingly measured the way that brilliant supernovae appear to get dimmer with cosmic distance. They find an excellent match to an accelerating expansion of the cosmos. It is a finding supported by a variety of astrophysical evidence, such as the growth of clusters and groups of galaxies across cosmic time, and the ethereal imprint of matter on the universe's all-pervasive radiation field—as I'll explain below. The bottom line is that about 5 billion years ago the universe stopped slowing down and started speeding up.

So what's causing this? The simplest answer is that we don't yet know, and, acknowledging our ignorance, we have labeled the cause of this all-pervasive acceleration with surprising honesty as "dark energy"—something we don't completely understand. One good possibility is that it is the energy of the vacuum of space itself, a seething ocean of "virtual" particle pairs popping in and out of existence that is a consequence of the deepest nature of quantum mechanical fuzziness and uncertainty. This ocean has peculiar properties, exerting a negative pressure—a repulsive gravitational field. An expanding universe simply makes room for more and more of this same "vacuum energy density," which comes to dominate the energy of the cosmos and pushes space ever farther apart.

Whatever dark energy turns out to be, right now it constitutes about 70 percent of the total matter and energy contents of the universe, and regardless of the details, it looks like it's here to stay. This turns out to have incredible implications for the period of cosmic time in which we find ourselves.

In 2007 the physicists Lawrence Krauss and Robert Scherrer presented a provocative study on this issue. Their focus was on the implications of an accelerating universe for the astronomical observations of any inhabitants, and specifically, how the universe of the future will present an entirely different face to wannabe cosmologists.

To do this, they imagined what the universe would look like to a species like us living in a galaxy somewhere in the cosmos 100 *billion* years in the future. Supposing that these creatures fabricate devices like telescopes to peer out into space, they will find that beyond the stars of their galaxy there is . . . nothing. Why? Because dark energy will have pushed the expansion of the universe to a point where visible light from other galaxies will have been stretched into effective invisibility. The cosmos beyond the galaxy will have faded out of sight.

Of course, these beings of the future won't necessarily be perturbed; they'll simply observe that the universe of visible light consists of their "island universe," their galaxy, and nothing else. It's that simple. But, as Krauss and Scherrer asked, given such a limited outlook, how will this species be able to develop an accurate theory of cosmology? Because it's not just the visible light of distant galaxies that dims into nothingness; other vital signatures of a universe with a finite age and a big bang will wash out as well.

In the 1960s scientists detected a pervasive cosmic background of microwave radiation, flooding space. This radiation was critical in proving the idea of a big bang. The microwave background is the remnant of the last epoch when the universe was hot enough to be opaque to light, some 380,000 years after the Big Bang and some 13.8 billion years before now. We can detect this remnant as an extremely smooth, but not quite perfectly uniform, hum of microwave photons racing through the universe in all directions. But 100 billion years in the future, the expansion of space will have reduced this background of microwaves to a trillionth of its present intensity, and stretched the photons into the ethereal realm of meter-length radio waves. Going even further into the future, someone looking from inside a galaxy will be unable to see even this, because local interstellar gas forms a near-impenetrable barrier to these ever-stretching electromagnetic wavelengths.

That's not all: the very balance of the raw cosmic mix of elements will have shifted in this future. Today, we can still see that hydrogen comprises roughly 74 percent of the mass of normal matter in the cosmos, and helium comprises 24 percent—a mix that is very close to the original, primordial proportions of hydrogen and helium. Together with tiny traces of deuterium (a heavy isotope of hydrogen), the balance of these elements is a critical clue to the hot, dense state of the young universe, a fingerprint of the Big Bang. But 100 billion years hence, stars will have converted more and more hydrogen into helium through nuclear fusion, shifting the latter's proportion toward a hefty 60 percent. The original proportions will be lost, and the traces of delicate deuterium that we can measure today will also have largely vanished, either destroyed by stars or lost from sight in the now-invisible spectrum of light from distant galaxies. Another smoking gun will have fizzled out.

In fact, the history of the stars themselves is at an interesting juncture now. Astronomers have known for the past twenty years that the overall rate at which stars have formed within galaxies was significantly higher in the past. Recent work, entailing heroic efforts of telescopic observation to map and characterize galaxies at many different cosmic ages, has helped pin down the details with unprecedented precision. It seems that more than half of all the stars we see today were produced between 11 billion and 8 billion years ago in a bacchanal of stellar production. Today's rate of stellar birth is barely 3 percent of what it was 11 billion years ago, and is declining quite rapidly. This means that for the rest of the universe's existence, only 5 percent more stars will ever be produced than have already been produced.

It's a disturbing thought. We exist at what you might call the beginning of a long cosmic twilight. And because small red stars are the most numerous stellar variety, and the longest-lived, the universe is effectively going to become fainter and redder, and it's going to be that way for a very long time. Many galaxies today produce hardly any new stars. Scientists believe that our Milky Way itself is in transition, shutting down its production of new stars and planets, making one or two stellar systems per year. It's somewhere in the middle of the range of today's productivity.

Why is this so? Partly it's because gas and dust from previous generations, the raw material for stars, gets spread out again after being initially drawn together by gravity. The energy of stars and supernovae, along with that produced by matter falling toward giant black holes, acts to disperse material in galaxies. And galaxies aren't growing and merging—processes that stir things up and can stimulate the condensation of new stars from the interstellar murk—nearly as much as they once did. Galactic mergers can still happen, though. In 4 or 5 billion years' time our neighboring galaxy Andromeda will come lumbering into us in a grand act of cosmic collision that will likely set off a round of new stellar production. It won't last very long on the cosmic timescale, however, maybe a couple hundred million years or so, and afterward the biggest and brightest of these new stars will die and we'll return to an inevitable dim red future.

Surveying these facts, we're led to the conclusion that we exist during what may be the only cosmic period when the universe's nature can be correctly inferred by observing what is around us. Ten billion years ago, when the universe was between 3 and 4 billion years old, we would have struggled to detect the emergence of dark energy and its influence on the expansion of space. In another 100 billion years, observers of the cosmos are likely to deduce that they live within a static universe. Few stars and planets will come and go in birth and death, but there is no easy way to recognize that space beyond the galaxy is expanding, and no easy way to deduce that the universe has a finite age.

All of this is intriguing, but there is one more issue that is most critical of all. Do we in fact know that the universe we observe around us today tells the whole story? What if we are just like those slightly hapless denizens of the far, far future—our vision of reality obscured by the nature of the universe itself? I don't think any of us have the answer to that, but it does highlight yet another challenge in our quest to understand our cosmic significance. Just as our celestial circumstances have impacted our scientific progress, as we saw, for instance, in the case of the shape of Mars's orbit when Kepler lived, our assessments of our place in the cosmos are powerfully influenced by what we know about the age and size of the universe. In this vision of the distant, isolated future, the inhabitants of a galaxy would realize that their sun is but one of a few hundred billion— much like our own circumstances in the Milky Way. But this is the sum total of the playing field before them, the totality of the cosmos, and it's very different and much more diminutive than the cosmos we know ourselves to be in today.

They might figure out that this modest universe of theirs has not been around forever, by noticing that as time goes by more and more hydrogen is converted into heavier elements. Turning the clock back would suggest an era in the past with no heavy elements, and for that to occur, there must have been no stars before a certain time. With enough stellar archaeology and astrophysics they could deduce that the oldest small red stars and stellar remains were indeed close to a hundred billion years old. I don't know what cosmology they could construct to explain these observations, but I'm sure it could appear quite logical. However, their cosmos would be small—a tiny universe compared with the one we know, a finite one in terms of stars and planets and opportunities for life. It would also be ancient by the clock of stellar astrophysics.

What answers about their cosmic significance would beings find in such a place? The challenge facing this hypothetical future species may not be so different from the one we face. There may be vital pieces of information missing from our view of nature, too, pieces that we don't even realize we're missing. Recognizing this, we should be ready to try some new strategies, to push beyond cosmic minutiae, rare Earths, biological dice-rolling, and the challenges of post hoc statistics. The picture of our cosmic mediocrity still stands pretty strong, but so too do some unusual aspects of our place in the universe. It looks like we'd better get ready to get our hands dirty as we dig into the uncomfortable truths about the evidence that surrounds us.

8

(IN)SIGNIFICANCE

We all reside on a small planet orbiting a single middle-aged star that is one of some 200 billion stars in the great swirl of matter that comprises the Milky Way galaxy. Our galaxy is but one of an estimated several hundred billion such structures in the observable universe—a volume that now stretches in all directions from us for more than 270,000,000,000,000,000,000,000 (2.7×10^{23}) miles. This region has grown to such a scale because of the continuous expansion of space that began some 13.8 billion years ago with the Big Bang. Astronomers have estimated that at least a billion trillion stars occupy this gaping abyss, and that many more have come and gone in the past billions of years.

By any paltry human standard this is a lot of stuff, and an awfully large amount of room. Our species has sprung into existence within the barest instant of this universe's long span of history, and it looks like there will be an even longer future that may or may not contain us. What can our significance possibly be? The quest to try to find our place, to discover our relevance, can seem like a monumental joke. We must be appallingly silly to imagine we can find any importance for ourselves at all.

Yet here we are trying to do just that, despite the mediocrity implied by the Copernican Principle, which has been one of our greatest guides for the past few hundred years. It's served as a major signpost on our journey to discern the underlying structure of the cosmos and the nature of reality. But in the course of this book we've seen the growing quantitative evidence that we face a conundrum in understanding our significance. Some discoveries and theories suggest life could easily be ordinary and common, and others suggest the opposite. I think our quest is beginning to produce some answers, and we're not crazy to imagine discovering our cosmic status.

So how do we get to the crux of it? How do we begin to pull together all of these strands of discovery, observation, and hypothesis—from bacteria to Big Bang—to explain whether or not we're special? For that matter, do these particular threads really connect into a single coherent picture? Or are some strands more important than others, or do they even contradict one another? Maybe, for instance, the precise architecture of our solar system is not as important for the origin and evolution of life here as we think, or perhaps it obscures something going on at a deeper level in the cosmic environment? And as we learn more about both the macrocosm and the microcosm, what does it all imply for our efforts to find out if there are other living things out there, and how do we take the next steps? Take a deep breath: we're about to tackle the fundamental nature of life itself.

❧

I began this book with the story of Antony van Leeuwenhoek peering into the alien world of the microcosm. In that remarkable descent, sliding down the ladder of physical dimensions into the thriving universe within us, was one of the first clues that the components of our bodies, our arrays of molecular structures, exist at one extreme end of a spectrum of biological scales. Until van Leeuwenhoek's moment of surprise, I doubt that humans had the opportunity to think about this fact in anything more than a superficial way.

There are organisms on Earth that are physically larger and more massive than we are—just look at whales and trees. There are also tightly knit collaborative ecosystems that we could argue are the largest living things of all: the forest honey fungus *Armillaria*, which as a cloned collective can span a couple of kilometers on a side, for one example. Yet we are much closer to this upper limit of scale (a thousand times smaller) than we are to the microscopic end of life's spectrum. A great physical gulf separates us from the microcosm. The tiniest reproducing bacteria are a couple of hundred *billionths* of a meter across; the littlest viruses are ten times smaller than that. The human body is roughly ten million to a hundred million times larger than the simplest life we know of.

Among warm-blooded terrestrial mammals we're also on the large side, but not quite at the extreme top. At the opposite end of the scale are the pygmy shrews, diminutive scraps of fur and flesh barely two grams in weight. They exist at the edge of feasibility, their bodies endlessly leaking heat that they can barely compensate for by voracious eating. But most mammals are closer to this size than to our size: so much so that the global average body weight of the mammalian population is forty grams, or less than two ounces. The reason for our largeness may be evolutionary drift: a successful niche can subtly encourage organisms to bulk up.

It's an undeniable observation that we exist at this border, this interface between the complex diversity of the biologically small and the limited options for the biologically large. Consider, too, our planetary system. We've seen that it's unusual in certain respects. The Sun is not one of the most numerous types of star, our orbits are at present more circular and rather more widely spaced apart than most, and we don't count a super-Earth among our planetary neighbors. If you were an architect of planetary systems, you would consider ours to be an outlier, a little bit off from the norm. Some of these characteristics stem from the fact that our solar system has escaped wholesale dynamical rearrangement, compared with the majority of planetary systems. This does not mean that we're assured a quiet and peaceful future—we've seen that a few hundred million years along, a more chaotic period could overtake our system. And another 5 billion years into the future the Sun will inflate with the onset of a spasmodic old age and drastically revise the properties of its array of planets. All indications are that today we also live at an interface or border in time, a transition between a period of stellar and planetary youth and one of encroaching decrepitude. Our existence in this period of relative calm is, in retrospect, not so surprising. As with so many other aspects of our circumstances, we live in a temperate place, not too hot or cold, not too chemically caustic or chemically inert, neither too unsettled nor too unchanging.

We've also seen that this astrophysically calm neighborhood extends well beyond our local galaxy. In terms of the universe as a whole, we exist in a period many times more ancient than the fast tumult of the young, hot cosmos. Everywhere the production of stars is slowing down. Other suns, and their planets, are forming at an average rate that is barely 3 percent of that 11 billion to 8 billion years ago. The stars are slowly beginning to go out across the universe. And in grand cosmological terms, only 5 or 6 billion years ago the universe was decelerating from the Big Bang. Now we're again in a period of gentle transition. Dark energy, stemming from the vacuum itself, is accelerating the growth of space, helping to quash the development of larger cosmic structures. But this means that life is ultimately condemned to a distant future of bleak isolation within an increasingly indecipherable universe.

Put all these factors together, and it's clear that our view of our inner and outer cosmos is highly constrained. It is a view from a narrow perch. Indeed, our basic intuition for random events and our scientific development of statistical inference might have been different under other circumstances of order or disorder, space and time. And the very fact that we are far isolated from any other life in the cosmos—to the extent that we haven't spotted or stumbled across it yet—has a profound impact on the conclusions we can draw.

Finally, to come full circle to the anthropic ideas we explored at the very outset, even the underlying properties of the universe suggest that it is finely balanced, near a boundary. A little too far to either side and the nature of the cosmos would be radically different. Tweak the relative strength of gravity and either no stars form, no heavy elements are forged—or huge stars form and are quickly gone, leaving nothing of any import in their wake, no descendants. Similarly, alter the electromagnetic force and the chemical bonds between atoms would be too weak or too strong to build the diversity of molecular structures that allows such incredible complexity in the cosmos.

•

I'd argue that these facts are pushing us toward a new scientific idea about our place in the cosmos, a departure from both the Copernican and anthropic principles, and I think it's well along the road to becoming a principle in its own right. Perhaps we could call it a "cosmo-chaotic principle," the place between order (from the original Greek *kósmos*, meaning a well-ordered system) and chaos. Its essence is that life, and specifically life like that on Earth, will always inhabit the border or interface between zones defined by such characteristics as energy, location, scale, time, order and disorder. Factors such as the stability or chaos of planetary orbits, or the variations of climate and geophysics on a planet, are direct manifestations of these characteristics. Too far away from such borders, in either direction, and the balance for life tips toward a hostile state. Life like us requires the right mix of ingredients, of calm and chaos—the right yin and yang.

Proximity to these edges keeps change and variation within reach, but not so close that they overwhelm a system constantly. There are obvious parallels to the concept of a Goldilocks zone, which proposes that a temperate cosmic environment for a planet around a star exists within a narrow range of parameters. But for the existence of life, the hospitable zone may be much more dynamic—it need not be fixed in space or time. Rather, it is a constantly drifting, twisting, flexing, multiparameter quantity, like the paths traced by a dancer's limbs.

If it's a universal rule that life exists only under these circumstances, it raises some intriguing possibilities about our cosmic significance. Unlike strict Copernican ideas, which stress our mediocrity and therefore suggest an abundance of similar circumstances across the cosmos, the notion that life requires a varying and dynamic alignment of parameters narrows the options. The opportunities for life implied by this new view also differ from anthropic ideas, which at their most extreme predict as little as one sole occurrence of life across all space and time. Instead, this new rule actually identifies the places where life should occur, and the potential frequency with which it does. It specifies the fundamental characteristics necessary for life within a virtual space of many waltzing parameters—it maps out the fertile zones.

Such a rule about life does not necessarily make living things some special part of reality. Biology may be the most complicated physical phenomenon in this universe—or in any amenable universe. But that's possibly as special as it gets: a particularly intricate natural structure that arises under the right circumstances, between order and chaos.

Several people who are studying the biological universe have suggested we adopt this dynamic way of conceptualizing life, as a phenomenon hovering on the brink of disorder, or on the edge of order. I remember a conversation years ago with the pioneering astrobiologist and physicist Michael Storrie-Lombardi in which he expressed this idea: life is something that happens on the edge, wherever that edge appears. By this he meant that life is a collection of phenomena at the boundary between order and chaos. Across that interface we can imagine that there is something akin to a voltage difference, a gradient of electrical potential that can be tapped into to drive a current. Except this biological gradient is multidimensional, an intersection of available energy, order and disorder, and time.

Others have come to similar conclusions. The theoretical biologist Stuart Kauffman, who studies the nature of complexity itself, has suggested that intricately structured systems of biology can emerge spontaneously from the combined effects of many simple rules or laws. Altogether, these simple rules and behaviors—of atoms, molecules, and thermodynamic systems—can produce enormous complexity and chaos, yet unexpected structures will appear out of this mess and "self-organize" into what is effectively something new.

At the same time, we're beginning to scope out the properties of the places in our universe where emergence happens, the borders between states of matter, space, and time—from galaxies to gas, stars, and planets. It is fascinating how that cosmic journey is arriving at precisely the same interpretation—that these edges and interfaces are the places where life occurs. And this conceptualization of where life fits into the grand scheme of nature leads directly to a way to resolve the conundrum between the persuasive, but unresolved, arguments that life must be abundant and that it is exquisitely rare.

•

In this book I've shown that an array of observations about chemistry, biology, and planets are telling us that the mechanics of life are an unsurprising extension of what we know about the universe. The elemental and chemical nature of the cosmos produces the necessary building blocks that go into life on Earth. And the basic underlying processes that life operates by—interleaved and interlocked metabolic processes carried through space and time by microbial organisms—piggybacks onto this same chemical bedrock.

There is no apparent peculiarity about life on Earth in that sense. The raw materials are everywhere, from interstellar space to protoplanetary systems, and preserved in the primitive leftover meteoritic and cometary material of our solar system. Furthermore, what we know about planet formation suggests mechanisms that could readily set the conditions for life's origins on a young rocky planet. Again, there is no obvious barrier between the common contents of the universe and the molecular and thermodynamic components of life on a planet such as Earth. The icing on the cake is the sheer abundance of rocky planets that we are now confident exist in our galaxy, tens of billions at present estimates, many of them pointing toward a wealth of conditions suitable for life. In fact, the evidence clearly describes a decentralized set of conditions that Copernicus would be proud of.

Notably, if life is rare, it is rather striking that the universe is nonetheless so good at setting the right stage. It need not have been this way. Even anthropic arguments only require that life is *possible*, not that it fits so neatly into the cosmos. If such unrealized potential is the case, it means that there really is something "special" that enables the transition from abiotic chemistry to biotic chemistry, something that could occur only in places identical to the Earth—a proposition that I've argued has little statistical weight at present.

However, like it or not, we also have conflicting observations about our place in the universe. We know from our galactic surveys that the Sun is not the most common type of star. We know from our discovery of exoplanets that our planetary system is not the most common arrangement of orbits and spacings. The solar system doesn't even contain representatives of the most commonly occurring types of planets, and it seems to have escaped some of the dramatic rearrangements that a majority of systems must go through. That doesn't mean it's invulnerable to the same long-term orbital chaos that plagues all planetary systems, but it's less prone to destructive change than many systems are.

We also exist at one of the few periods in the history of the universe when our eyes and telescopes have the opportunity to make meaningful observations about the nature of what surrounds us. If we lived deeper in the past or further in the future, we'd be missing out on vital information. On a more local level, we also exist in a celestial environment that neither hides the nature of the universe from us nor makes it particularly easy to deduce. Other places could present a much more readily interpreted picture of the structure of the cosmos and the characteristics of fundamental laws such as those of gravity and mechanics.

Looking at our planetary environment we can also, if we choose, find evidence that our existence as complex-celled and intelligent organisms is contingent on the alignment of multiple phenomena. Many of these alignments seem to be sheer chance—the result of mass extinctions or swings in environment brought on by all manner of forces, some external to the Earth, like giant dinosaur-killing asteroids. Other factors could include the fusion of two primitive life-forms (as in the case of mitochondria), taking a new, seemingly unlikely, and very necessary step toward complex life.

So are we unusual or not? Our powerful tools of mathematical probability, and the objective truths about the bias in retrospective interpretation of events, clearly indicate that neither side is yet a winner. But we are much, much closer to an answer than we have ever been in the history of the human species; we are on the cusp of knowing.

My own conclusion takes the strands of everything we've been discussing and weaves them together. Consider what I said about life's status as an emergent phenomenon, and how life happens at the swirling borders between an array of variable physical circumstances.

Now apply that rule to the conundrum we face between mediocrity and unusualness. What do we end up with?

We end up with this: Our place in the universe is special but *not* significant, unique but *not* exceptional. The Copernican Principle is both right and wrong, and it's time we acknowledge that fact.

Take a look at the evidence, from the chemistry of the cosmos to the dynamism of planet formation, and the joint evolution of biology and geophysics here on Earth. I think there is little doubt that a huge number of opportunities exist for living environments, all based on the same building blocks and principles. Because of this multiplicity, our specific human biology, its evolutionary history, and its connections to our planetary circumstances could well be unique—if measured with a fine-enough pair of calipers. But this does *not* have to imply that life—even complex life—can't reach similar states by following other pathways. We could be special yet surrounded by a universe of other equally complex, equally special life-forms that just took a different trajectory. Our uniqueness is tempered by being nothing exceptional in the panoply of life; we are just one representative of the phenomenon.

Now, in any post hoc analysis of a phenomenon, contrary to intuition, the correct default position is to assume that it represents the most common type of outcome. (That baseball that hit Joe was bound to hit someone.) That's pretty clear. So it could be that the occurrence of life like that on Earth is independent of the finer details of a celestial environment—in which case, all the unusualness of our solar system is just a red herring.

Alternatively, at the opposite extreme, it could be that certain aspects of that local environment are absolutely essential—a parochial fine-tuning for life. But as I've shown, evidence that seems to point to that latter condition may be very misleading. And so I find myself aiming for this position of our specialness, or near uniqueness, but not significance. Celestial conditions are going to lead to lots of worlds more similar to the Earth than not. Whether they're a bit smaller or larger in size, the potential is there. We already know that there are tens of billions of rocky worlds in our galaxy. None will be a *precise* match to the Earth's present, past, or future state—they simply can't be because of randomness and chaos. But to my mind this diversity doesn't have to pose a problem. If environmental differences are modest, simple and complex life could find a way to emerge.

One underlying assumption here is that the basic mechanics of living organisms can be cobbled together in a variety of ways from the same building blocks. That's effectively saying the division of life on Earth into the great domains of bacteria, archaea, and eukarya is just one outcome, one option. However, some scientists argue for what's called convergent evolution, the idea that there are only a finite number of useful biological blueprints, and that evolution will always drift toward them. For complex organisms, this argument has been used to help explain how similar "camera eyes" exist in both vertebrates (like humans) and cephalopods (like squid), even though we and they are on evolutionary tracks that parted a very long time ago.

The principle of convergent evolution has also been used to argue that there are a limited number of "useful" protein behaviors, a limited mix of different molecular structures that can perform the same functions. This finite protein toolbox suggests that the same molecules must occur anywhere in the universe in order for life to work. Maybe such a biochemical homogeneity reduces the number of possible biochemical mechanisms or biological blueprints for life across the universe. But I'm not convinced we know that for sure, and for the same reasons that make it so hard to retrospectively evaluate random events: to use the Earth as a template risks being grossly misled.

What I'm describing here is, I suppose, the most optimistic interpretation of the present evidence. It allows for an abundance of life as well as our specialness. It's consistent with what statistical evaluations can tell us as the moment. It also has the marvelous property of being testable, and leads to what I think is the most intriguing possibility—that we might forcibly transcend our circumstances and become not just unique, but significant as well. Because as much as the hypothesis I've presented is the end result of carefully weighing the growing body of evidence we have available to us, the quest is far from over. The discoveries and ideas you've read about here are leading us to new territory. This frontier is the setting for the final stories that I want to leave you with. Some are about risky scientific business, some are blue-sky propositions that I want to share, and some are about questions we all need to ask ourselves.

※

On August 18, 1977, the American astronomer Jerry Ehman sat at his kitchen table flipping through sheet after sheet of computer-generated printout. Raining down these pages was a cryptic stream of blank spaces and inky digits falling into regularly spaced columns.

As Ehman looked carefully through this forest of information, his eyes lit upon a very unusual feature on one page. Instead of its usual output of low numbers, the computer had printed a column that read, from top to bottom: "6EQUJ5." Grabbing a red pen, he circled this group of characters, and to the left, in the margin, he wrote "Wow!"

This small section of paper with its imperfectly printed characters, together with Ehman's emphatic note, represent what some people think remains the best evidence of a signal from the cosmic depths that was of artificial, purposeful, and intelligent origin.

A few days earlier, on August 15, 1977, this printout had spilled forth from an analysis of radio signals detected by the Big Ear telescope in a field outside Delaware, Ohio. Big Ear was a rectangular structure, larger than three football fields, covering the ground in metal paneling and bookended with two sloping fence-like structures. At this time, it was purposely listening for something very specific.

As the Earth rotated and the sky drifted across its view, Big Ear was catching radio signals in a set of fifty distinct frequency channels. These included some that overlapped with a special natural frequency—the frequency at which atoms of hydrogen emit radiation when their proton and electron flip between quantum spin states.

This may not sound very exciting, but for scientists, this frequency

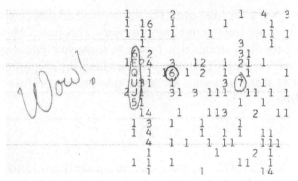

Figure 15: The "Wow!" signal, an unrepeated burst of radio power from beyond the Earth (J. Ehman, and Big Ear Radio Observatory and North American Astrophysical Observatory)

(known as the 1400 MHz or 21-centimeter line) is important. It reveals the glow of interstellar and intergalactic hydrogen gas, and it can reveal the moisture content of our atmosphere and even the salinity of our oceans here on Earth when detected from space. This frequency also sits at an especially quiet spot within the galactic hubbub of radio waves, an attractive place to gather to listen for interesting phenomena. For this reason it's often termed "the cosmic watering hole" in the electromagnetic spectrum.

So it's a special frequency, universal in nature, but it's also a frequency that shouldn't typically flash on or off, or do anything more complicated than gently sit there humming across the cosmos. And this was just why Big Ear was listening for it, because in August 1977 Jerry Ehman and his colleagues were engaged in a search for extraterrestrial intelligence, more commonly known by its acronym: SETI.

"6EQUJ5" signified a sudden pulse of radio energy when it appeared on Big Ear's printout. Usually the faint signals of natural noise only rated blank spaces, or digits such as 1, 2, or 3. But if the signals got strong enough the computer would have to shift up to letters—and by the time it got to "U" it meant a signal about thirty times more powerful than the cosmic background. This pulse lasted for the duration of Big Ear's attention span on any one spot on the sky: seventy-two seconds. It also came in at almost exactly the atomic hydrogen frequency, the cosmic watering hole. But then it was gone. And it didn't come back—ever.

Much has been written about this "Wow" signal. Jerry Ehman himself has carefully evaluated many mundane possibilities for its origins, but most come up blank. It's very unlikely to have been something on Earth, or even in Earth's orbit—a passing satellite or a space mission. But if it was cosmic in origin, we simply don't know what it was, or even where it came from, because Big Ear wasn't capable of pinpointing a location to any great precision.

Since the 1970s astronomers have learned a lot more about the so-called transient universe, natural phenomena such as gamma-ray bursts, pulsar glitches, burping black holes, and other events that come and go. Yet no one example is a clear match to what Big Ear caught a glimpse of, and so the mystery remains.

It's a terrific story, but it also highlights one of the intrinsic difficulties that SETI grapples with—the problem of confirming and interpreting fragmentary and fleeting information. Indeed, our effort to directly detect purposeful signs of other intelligent life and civilizations in the universe has to date come up empty-handed.

The absence of indisputable evidence for extraterrestrial intelligence has provided endless fodder for speculation, the most sophisticated of which centers on what's known as the Fermi paradox, after the famous Italian physicist Enrico Fermi. The story goes that in 1950, over lunch with colleagues, Fermi pointed out that the galaxy is old enough, and stars are numerous enough, that if life is reasonably common there should already be advanced civilizations buzzing around every corner of space. The paradoxical question he raised was why we hadn't seen them.

On the face of it this is a most excellent question, and entire books have been written on the subject. The problem with resolving the paradox is, again, a lack of information. We can think of innumerable reasons why no one has shown up on our celestial doorstep to say hello: interstellar travel may be very difficult, intelligent life may self-destruct, maybe life isn't common at all, maybe it's just too alien, perhaps they choose to remain silent, or they're already here but we just don't realize it. You can insert the joke of your choosing at any point.

Any real evidence would break the impasse. So while the direct search for extraterrestrial intelligence is a challenging, risky, and problematic thing to pursue, I fully support the effort. In the absence of knowledge, the only thing you can do is to try. And this is the vital point. Again and again in this book we've come up against the need to determine the next step, an actual successful test for whether or not life exists beyond the Earth. SETI represents one extreme, the whole hog. But there are other options.

For example, the advent of exoplanetary science has encouraged a new search strategy for finding life. This strategy looks not for structured signals or artificially produced phenomena, but rather for evidence of the same interwoven biogeochemical mechanisms that have been at play here on Earth for the past 4 billion years.

Life alters and manipulates the chemistry of an environment, pushing it out of equilibrium. Gaze at the Earth from afar with the right instruments, for example, and you might detect the presence of both oxygen and methane in its atmosphere. This is a peculiar combination. Oxygen is highly reactive, and over time it should combine with the minerals on the surface of a rocky planet, taking it out of the atmosphere. Oxygen reacts even more vigorously with methane, combining into carbon dioxide and water. Detecting both of these gases in an atmosphere tells us that something must be continually replenishing them, and one of the best sources is life itself.

Other molecules present potential biosignatures, detectable in the spectra of light absorbed or emitted by the contents of another world. Gases such as nitrous oxide and sulfur compounds can be participants in planetwide metabolic processes, and other physical aspects of an Earth-like world can reveal interesting clues to what's going on at a local level. The glinting reflection of light from oceans, the coverage and texture of water vapor clouds, and even the tell-tale colors of photosynthetic pigments all offer signposts to what's happening on the ground. Take terrestrial plants, for example. The chlorophyll in their leaves (contained in chloroplasts that probably started out as endosymbiotic cyanobacteria) absorbs many frequencies of visible light, but reflects green wavelengths, making them appear green to the eye. But plants also strongly reflect and transmit near-infrared light, causing a remarkable ten times more infrared radiation to bounce off them than visible radiation. We exploit this in our satellite surveys of the Earth, using it to clearly map out vegetation and vegetation loss. Exactly how plant life performs this optical trick seems to involve both internal cellular structures and their photosynthetic pigments. It could be an Earth-specific phenomenon, but it could also be a characteristic of any biosphere that soaks up stellar radiation to help power itself.

Phenomena like these offer hope that as we get better and better at capturing the light from distant worlds, or disentangling their atmospheric contents using the backlight of a star, we may spot these biosignatures. Life's exuberance often leaves a filthy fingerprint. It's difficult to find, for the same reason that detecting planets in the first place is difficult: planets are faint and stars are bright. Nonetheless, the near future of astronomical technology is going to provide this opportunity for at least a handful of planetary systems that are close enough to us for telescopes to gather sufficient light.

And this brings us to another sixty-four-quadrillion-dollar question, one that I brought up right at the start of this long tale. Is the actual propensity of this universe for abiogenesis, the measurable abundance of life, a litmus test—a new way to probe the deepest fundamental physical laws and natural constants, and to in turn evaluate the significance of life? Notice that this is more elaborate than an anthropic or fine-tuning test, which posits that the cosmos must simply meet certain fixed criteria for life like us to occur. In those formulations the answer is essentially a binary option: life or no life. Chances are that the real answer is more like a "figure of merit," as they say in engineering—a sliding scale, a measure of the fertility of the cosmos.

That fertility may well be the key to linking physics and life, but we don't yet understand enough to know what the properties are that drive the specific richness of life in this universe. However, there could be a way to figure this out. A part of the challenge is to separate our own local circumstances from the underlying parameters that govern the universe. For example, something as simple as the age of the cosmos obviously influences the options for more or less life. We can see that life as we know it had to be absent until the first stars had formed the first heavy elements. In fact, it probably took a few stellar generations before sufficient elements existed to even form rocky planets. We can also imagine that in the distant future of dimly lit and isolated galaxies of low-mass stars, conditions may be less amenable for the emergence of life. Aging rock-rich planets will have less and less geophysical activity to keep recycling their surface chemistry.

There must be other properties that help set the probability level of life occurring at any given cosmic time as well. Much like the fine-tuning parameters in an anthropic view of the universe, these could be quantities such as the strength of gravity or the chances of atoms and molecules forming, and the deeper physics that gives rise to these properties. Such factors ultimately help determine the production of stars and planets and their subsequent evolution, as well as the details of biochemically friendly environments. These features must be intimately linked to the origins of life in the first place, as well as to its ability to flourish. If we could write a recipe for those origins, I think we'd have the answer. We'd implicitly know how the cosmic parameters determine the abundance of life at any point in the history of the universe. But is there a precise recipe at all?

I've already suggested, like many other scientists, that life is much more of an emergent property—something that springs from an essentially impenetrable dance of "nonlinear" interactions and behaviors stemming from simpler rules—part of a cosmo-chaotic principle. Those rules are the physical underpinnings—from molecular bonds, to the deeper symmetries of subatomic particles, to the dimensional nature of reality—but it's hard to gauge their precise contributions to the recipe. That's because the tangled way in which those rules interact is itself a nonlinear function of those rules! In other words, the individual influence of each of those properties may become indecipherable—much like trying to deduce the rudimentary principles of thermodynamics using only measurements of weather and climate on Earth. The system's innate sensitivity to initial conditions may obscure root causes and effects in the end result.

You may guess where I'm going with this. There's something awfully familiar about the sound of this problem, and it's the sound of chaos theory. It's a lot like the challenge we encountered in understanding planetary orbital dynamics and the long-term stability or instability of our solar system. You'll recall that in planetary systems there are simple rules as well, but complex nonlinear interactions produce an array of past and future trajectories—a huge bundle of pathways and possibilities. To find out what happens if you tweak the rules, you need to follow countless routes, each careening off from different starting points to unpredictable outcomes.

To understand the frequency with which a cosmic setting produces life, we'll have to perform a similar experiment. We'll have to simulate the conditions produced by a range of cosmic properties, to see how well and how often they generate the complex phenomena out of which life emerges—how many viable trajectories there are. We'll also have to apply our Bayesian skills to weigh the possibilities, to express our honesty about our ignorance of the deeper physics of reality.

It's not hard to see that this is an overwhelming computational and theoretical challenge. It parallels another seemingly impenetrable and unnerving question: understanding the human mind. In the recent past, scientists have argued that it's in principle possible to build a simulation of a mind, a true artificial intelligence, once we develop computer software that's sophisticated enough to mimic digitally our tens of billions of neurons. However, some researchers, like the English scientist Roger Penrose, have argued that there are deep relationships to the quantum world that play vital roles in minds and consciousness, impossible to capture in digital code. It might be that the only way to simulate a mind is to actually build one, a structure full of the same messy chemistry and biology as ours. Only such a simulacrum would have the computational capacity and natural trickery necessary to match what evolution has wrought over billions of years.

Perhaps we could do this more easily for life in the broader sense. We're already taking small steps toward constructing artificial microbial life-forms, cobbled together out of spare parts and lab-built DNA. But clearly, when it comes to the rule set, we can't alter the fundamental physics of these biosimulations, we can't play around with the cosmic underpinnings, and this presents a bit of a roadblock. Is the phenomenon of life in the cosmos something that we'll ultimately have to just accept and study without hope of the equivalent of a physicist's "theory of everything"?

I hope not. I think that we might be able to do better at simulating the trajectories of life at different cosmic parameter settings than some of the above ideas suggest. I am optimistic about this in part because our technological prowess is continuing to accelerate at an astonishing rate. We are learning unprecedented ways to manipulate matter at both an atomic and a subatomic level. Laboratory-bench physics now lets us tinker with the innate weirdness of quantum mechanics, to exploit its rules to build the most unexpected of things—from rudimentary quantum computers to fiber-optic simulations of black holes' event horizons, those gravitational points of no return from which even light cannot escape. There may well be a convergence of tools and techniques in our not-too-distant future that catapults today's impossibility well into the realm of the possible.

We have another potential strategy in our toolbox, too. And that is to get out there and start adding up the instances of life in the cosmos, if we can find them. The universe is the ultimate experiment. It also has a very useful and special property: it is big enough so that widely separated locations in space have already been effectively isolated from each other since before discernible atoms of matter existed.

In effect, every large chunk of the cosmos serves as an independent petri dish. It's a fact that cosmologists and astronomers make good use of for analyzing the properties of stars and galaxies as they evolve through cosmic time. Objects at the center of any big-enough region of the universe have never been directly influenced by, or related to, objects at the center of other, separate big regions. Each is in effect a unique island that has developed along its own trajectory, but governed by the same universal physical laws as all other islands. Ironically, this is an extension of the Copernican Principle: no place in the universe is exceptional, although it may have turned out slightly different than others.

We can play the same game in the quest for life. However, our solar system may be too small to provide more than one petri dish. Its planets are prone to cross-contamination of chemistry and organisms as asteroid impacts eject material across interplanetary space. A better option would be to do this star by star, but as we've seen, the transfer of material across interstellar space can also contaminate things. A more certain bet would be to divide up a large galaxy like the Milky Way, each zone representing a potentially independent sampling of the multitude of pathways that life might take. We could go further, reaching out to intergalactic space and treating entire galaxies as independent experimental incubators. If we could identify and quantify the nature of any life in these places, we could gather that giant map of trajectories and then look to see what universal cosmic underpinnings drive this sprawl.

The funny thing is that we already know that this sort of approach works in science, and for that we owe a direct debt to Antony van Leeuwenhoek sitting in his room in Delft in 1674. When van Leeuwenhoek saw microscopic organisms inhabiting every drop of water, and every orifice and excretion of humans and animals, he unwittingly drew up a blueprint for the exploration of life's hiding places. Today, scientists take the procedure of controlled sampling of microbial life for granted. To identify new species in the extreme environments of subterranean pockets of water, or deep beneath Antarctic ice, for example, researchers go to enormous lengths to isolate uncontaminated samples. Pristine micro-ecosystems can harbor organisms that have evolved by themselves for thousands, perhaps even millions of years, living cut off from the rest of the world. By looking in at these lonely microcosms we can learn a great deal more about the incredible biological strategies that develop, and we can probe the underlying biological principles at play.

To do the same for the cosmos is a wildly ambitious and optimistic idea. But the ultimate reward would be worth it. Back in the first chapter, I briefly discussed the scientific idea of the multiverse, a far-reaching way to account for the seeming coincidence of cosmic fine-tuning and life. By using life as a litmus test, we could probe this theory. Suppose we could determine the values or forms of physical constants and laws that set the level of life's likelihood and abundance in a universe. With that information we would conceivably be able to predict how much of the *multiverse* held life like us. In other words, we'd calculate our significance in the absolute totality of all possible realities.

That's quite a lofty ambition. To realize it, we're going to have to confront our Copernicus complex. We are, I think, still unlikely to be central to the universe, either astrophysically or metaphysically. But this does not preclude the possibility that the pathway of emergence that produced us is unusual in its details. We need to get comfortable with that degree of specialness, because it influences our outlook and our scientific strategies for reaching out to the universe. We can make that journey safely through our telescopes, or we can set out with a much bolder goal. I don't consider this goal a fantasy. It may be the most important choice our species ever makes, and it starts and ends with two questions.

❧

Can we ever transcend the cosmic circumstances of our existence? And do we want to remain special but insignificant?

The rules of engagement with these two challenges are a little unfair. If life always and without exception inhabits the border between order and chaos, it would appear that intentional cosmic growth requires extraordinary nimbleness. It's like the task facing an expert surfer trying to stay in place on the slippery and changing face of a great wave, a tube in space and time that will run out only to be replaced by the one behind it.

But metaphysics be damned. We know where we are, and we know what we need to survive (even if we don't always seem to know). We have emerged on a planet from inauspicious beginnings as microbes nearly 4 billion years ago. We've not only gained a conscious appreciation for this fact; we've managed to gauge the origins and contents of the universe around us. And we've found that there are tens of billions of other worlds out there, and enormously rich resources elsewhere in our own solar system.

So here we are at the next fork in the road, with a new kind of choice to be made. This challenge facing us is the remaining piece of the quest to learn our cosmic significance, and it reaches all the way to the foundations of our existence, as well as to our relationship to the mechanisms of natural selection and evolution. However modern humans began, with our remarkable brains and social structures, and however small our total population has been at times, there is no doubt that today we are a major force on this planet. Billions of us occupy the world, and even those remaining landscapes that we don't occupy in person, we've mostly altered by harvesting resources or subverting the environment. Whatever our intricate relationship is to the microbial overlords that help govern our habitat and our own biochemistry, we already stand out as something different among organisms.

Our reach deliberately extends far beyond the confines of this planet. For the past forty years the *Pioneer 10* and *Pioneer 11* spacecraft have been speeding away from us toward interstellar space. They are now some 10 and 8 *billion* miles from us, respectively. Launched only a few years after *10* and *11*, the probes *Voyager 1* and *Voyager 2* are pushing even farther out into the universe. *Voyager 1* is now more than 11 billion miles from us, more than 125 times farther from the Sun than Earth's orbit. It still communicates with us. Its faint whispers of radio telemetry tell us that it has reached a place where the pressure of our Sun's streaming particle radiation is giving way to that of the surrounding galactic space. This journey may have only just begun, but it was seeded long ago, when the first hominids trekked across the African savanna. To paraphrase the words of Carl Sagan, we were always wanderers.

Our true cosmic importance may ultimately spring from this same urge for expansion, a vital signature that natural selection has written in our human genes. It is who we are. It is part of what makes us special. And it is how we can, if we choose, *make* ourselves significant. The awesome barriers of interstellar distance and time, and the unsheathed forces of the cosmos, might forever impede us from spreading our fragile corporeal forms very far from our solar system, despite the fantasies we like to invent. But let's suppose we succeed in spotting the signs of life on another planet around another star in our galactic neighborhood. Even if the signatures are little more than chemical agents sensed in a spectrum of light, betraying the metabolic processes of microbial-style life, the possibility exists for more complex organisms. Someone else could be out there—alien, but within reach.

The discovery of that biological signature will mean a point of decision for us. We might not want to try sending ourselves on a trek to this other world that could last thousands or even tens of thousands of years. But we might consider building a representative. Whether this emissary is a sophisticated robot or a simple message carrier, its eventual arrival at another world would mark the only true opportunity to signify the fact that we once existed on a place special to us that we called, simply, Earth.

NOTES

3 *Antony van Leeuwenhoek stares*: A large amount of literature and number of resources exist on van Leeuwenhoek, sometimes known as the "Father of Microbiology." Although an amateur scientist in the sense that he had no formal training, he did become a member of England's Royal Society. Altogether he wrote more than five hundred letters to the Society and other scientific institutions describing his observations, which included some of the first observations of blood cells and sperm cells. An interesting bit of trivia is that in 1676 he was made trustee of the estate of Jan Vermeer, the great painter. Van Leeuwenhoek lived until 1723, dying at the age of ninety. An excellent resource is www.vanleeuwenhoek.com/.

4 *great work* Micrographia: The full title was "*Micrographia: or some physiological descriptions of minute bodies made by magnifying glasses with observations and inquiries thereupon*" (phew). Published in 1665 (London: J. Martyn and J. Allestry, first edition), it contained many, many drawings and discussions: "Of the Sting of a Bee; Of Peacocks Feathers; Of the Feet of Flyes, and other Insects; Of the Head of a Fly; Of the Teeth of a Snail; Of the Beard of a wild Oat; Of Diamonds in Flints; Of a Vegetable growing on blighted Leaves; Of a Crab-like Insect." Its impact was enormous. The English diarist Samuel Pepys wrote of it as "the most ingenious book that ever I read in my life." See also this short piece by P. Fara, "A Microscopic Reality Tale," *Nature* 459 (2009): 642–44.

4 *Robert Hooke*: The English polymath, born in 1635, died 1703, was an extraordinarily inventive person, and rose from comparatively humble beginnings. He became the "curator of experiments" at the fledgling Royal Society, and in addition to making advances in microscopy, he came very close to deducing some of the key elements of Newton's

law of gravitation. The biological term "cell" is attributed to Hooke, who first used it to describe the boxy appearance of plant cells under his microscopes.

5 *compound microscope*: Prior to van Leeuwenhoek's work, microscopes had been constructed using multiple lenses to magnify specimens—the simplest configuration being two lenses of different focal lengths, one at either end of a tube.

5 *made his microscopes*: Van Leeuwenhoek's techniques are still not entirely known. But by making tiny spherical lenses he was probably able to improve their overall optical quality, and avoid the need for careful polishing. Specimens in droplets of water could have in effect produced a tiny compound optic, with the water itself acting as a lens.

6 *well over two hundred:* Estimates vary, some sources refer to over five hundred—but this may have been the number of lenses, not the number of actual microscopes. Van Leeuwenhoek worked at this for some fifty years, so these numbers are probably not exaggerated.

6 *fateful drop of water*: Van Leeuwenhoek's notes seem to pinpoint this water as coming from Berkelse Mere, a small lake near Delft.

6 *Within these drops*: Van Leeuwenhoek wrote: "and seeing the water as above described, I took up a little of it in a glass phial; and examining this water next day, I found floating therein divers earthy particles, and some green streaks, spirally wound serpent-wise, and orderly arranged, after the manner of the copper or tin worms, which distillers use to cool their liquors as they distil over. The whole circumference of each of these streaks was about the thickness of a hair of one's head."

6 *of the human mouth*: Samples of human tartar went under the lens in 1683 and seem to have included the rod-shaped bacterial genus *Bacillus*.

7 *place in the universe*: Scientists were intrigued by the microscopic world, and observations of the reproduction of tiny organisms were counter to the prevailing notion of "spontaneous generation" at the time. Nonetheless, this discovery seems to have stirred far less controversy than observations of the larger universe did.

7 *till the mid-1800s*: Best known is the work of Louis Pasteur, who also firmly disproved ideas of spontaneous generation, and proposed that bacteria could not only spoil food but also cause human disease. Heating ("pasteurizing") food would help preserve it. Robert Koch demonstrated that the disease anthrax was caused by a bacterium.

1: THE COPERNICUS COMPLEX

11 *Aristarchus had just had a brilliant idea*: The original writings are lost. However, in *The Sand Reckoner* by Archimedes (in which he attempts

to calculate the number of grains of sand that fit into the universe), reference is made to Aristarchus's heliocentric idea: "His hypotheses are that the fixed stars and the Sun remain unmoved, that the Earth revolves about the Sun in the circumference of a circle, the Sun lying in the middle of the [Earth's] course, and that the sphere of the fixed stars, situated about the same center as the Sun, is so great that the circle in which he supposes the Earth to revolve bears such a proportion to the distance of the fixed stars as the center of the sphere bears to its surface." Sir Thomas Heath, *Aristarchus of Samos, the Ancient Copernicus: A History of Greek Astronomy to Aristarchus, together with Aristarchus's Treatise on the Sizes and Distances of the Sun and Moon: A New Greek Text with Translation and Notes* (Oxford: Clarendon Press, 1913), 302.

12 *approximately fifty-five*: The exact number of these crystalline spheres seems to have been either 47 or 55 depending on the version of Aristotle's model. Aristotle, *Metaphysics*, 1073b1–1074a13, in *The Basic Works of Aristotle*, ed. Richard McKeon (New York: Random House, 1941; The Modern Library, 2001), 882–83.

14 *geocentric "fix"*: Much of the detail in this chapter is derived from Thomas S. Kuhn's remarkable and extensive discussion in *The Copernican Revolution: Planetary Astronomy in the Development of Western Thought* (Cambridge/London: Harvard University Press, 1957; rev. ed., 1983), especially the deeper threads connecting this "cosmology" to the whole framework of scientific and religious thought across the ages.

14 *the Almagest*: A modern translation is *Ptolemy's Almagest*, G. J. Toomer (Princeton: Princeton University Press, 1998). The name derives from the Arabic, in turn derived from the ancient Greek for "greatest"; it is also known in Latin as the *Syntaxis mathematica*.

15 *little deviations here and there*: As we'll see, one of the problems was in the timing of planetary positions, and Ptolemy's model assumed all motion around epicycles and deferents was at constant velocity.

17 *seven critical and visionary axioms*: See, for example, André Goddu, *Copernicus and the Aristotelian Tradition: Education, Reading, and Philosophy in Copernicus's Path to Heliocentrism* (Leiden, Netherlands: Brill, 2010). An excellent history and discussion is in Owen Gingerich's *The Book Nobody Read: Chasing the Revolutions of Nicolaus Copernicus* (New York: Walker & Company, 2004).

18 *some historical intrigue*: Indeed, much has been written on what motivated Copernicus and what inhibited him. A fun, somewhat speculative take is Dava Sobel's *A More Perfect Heaven: How Copernicus Revolutionized The Cosmos* (New York: Walker & Company, 2011).

19 *Giordano Bruno*: See, for example, Ingrid D. Rowland's *Giordano Bruno: Philosopher/Heretic* (New York: Farrar, Straus and Giroux, 2008).

19 *Tycho Brahe*: A great deal has been written about Brahe, and for good reason: he was a colorful character with the means to pursue an interesting life. King Frederick II of Denmark also provided him with funds to help build an observatory, and gave him the little island of Hven in the Øresund near Copenhagen. This was where he built the observatory, Uraniborg—later expanded with an underground facility for greater stability. There were no telescopes, but structures and devices to measure precise positions and angular relationships of celestial objects using the human eye.

20 *witnessed a new star*: Brahe described his observations of what we now know was a supernova in *De Nova et Nullius Aevi Memoria Prius Visa Stella* ("On the New and Never Previously Seen Star") (Copenhagen, 1573). This, together with his observations of comets, led him to dispute the Aristotelian belief in an immutable cosmos.

21 *Kepler wound up inheriting*: An excellent resource that covers the development of Western astronomy and cosmology is by Arthur Koestler, *The Sleepwalkers: A History of Man's Changing Vision of the Universe* (London: Hutchinson, 1959; repr. Arkana / Penguin, 1989). In this book, Kepler is treated very much as the scientific hero of the times. Some accounts suggest that Brahe originally encouraged Kepler to work on Mars because it was confounding enough to keep Kepler out of Brahe's hair and to prevent Kepler from finding support for a Copernican system. But it seems that Kepler knew what he was doing. One can find a quote from Kepler himself in a letter of 1605: "*I confess that when Tycho died, I quickly took advantage of the absence, or lack of circumspection, of the heirs, by taking the observations under my care, or perhaps usurping them . . .*"

22 *conic sections*: These curves are, as the name suggests, literally the result of sectioning or slicing through a cone with a flat plane. Depending on the relative orientation, where the two meet can be described by a parabola, a hyperbola, an ellipse, or a circle.

22 *Galileo Galilei*: The Italian scientist used two lenses to fashion telescopes that produced a right-side-up image of distant objects. They were far from perfect in terms of optics, but his best yielded 30× magnification and could capture more light than the human eye alone. Like Brahe, Galileo witnessed a supernova (also seen by Kepler), and because he could see no parallax motion, decided it was a star and that the heavens were not immutable. His observations of three, and then four, moons moving around Jupiter led him to the insight that this confirmed the Copernican view: not all bodies move around the Earth.

25 *determinism in the universe*: Pierre-Simon Laplace felt strongly about this; he wrote in *A Philosophical Essay on Probabilities* in 1814: "We

may regard the present state of the universe as the effect of its past and the cause of its future. An intellect which at a certain moment would know all forces that set nature in motion, and all positions of all items of which nature is composed, if this intellect were also vast enough to submit these data to analysis, it would embrace in a single formula the movements of the greatest bodies of the universe and those of the tiniest atom; for such an intellect nothing would be uncertain and the future just like the past would be present before its eyes." Trans. F. W. Truscott and F. L. Emory (New York: Dover Publications, 1951), 4.

25 *Christiaan Huygens*: His thoughts on life in the universe were published posthumously in 1698 in *Cosmotheoros*.

26 *underappreciated scientific debate*: What has become known as the "nebular hypothesis" of the formation of the solar system from a cloud of rotating, orbiting, contracting material was probably first put forward in 1734 by Emanuel Swedenborg (yes, the theologian), further worked on by Immanuel Kant (yes, the philosopher) in 1755, and also described by Laplace in 1796. In these earlier versions the theory faced a huge problem because it could not readily explain why the planets carried 99 percent of the angular momentum of the system. It really wasn't until the early 1970s, when the Soviet scientist Victor Safronov produced convincing solutions for this and other issues, that the model came back into favor.

26 *Ceres and Vesta*: In today's nomenclature Ceres (590 miles average diameter) is considered a dwarf planet and Vesta (326 miles average diameter) is a minor planet.

26 *the stuff we now call helium*: Evidence appeared as a bright yellow "line" in the spectrum of the Sun's light, first seen in 1868. By 1895 helium was isolated from minerals on Earth.

27 *the so-called* Cosmological Principle: Strictly speaking, this was the *modern* cosmological principle. The underlying ideas can be traced back to Newton. In the 1920s both Alexander Friedmann and, independently, Georges Lemaître (the first person to propose the expansion of the universe) solved the equations of general relativity to determine the dynamics of a universe that was both homogeneous and isotropic. Later Howard Robertson and Arthur Walker did the same, leading to what is now known as the Friedmann-Lemaître-Robertson-Walker metric—in essence a matrix describing the relationship of space and time coordinates in the universe.

28 *Hermann Bondi*: The Anglo-Austrian Bondi (1919–2005) worked with Fred Hoyle and Thomas Gold to produce the steady-state theory of cosmology in 1948, and made a number of major contributions to relativity and astrophysics. The Copernican Principle comes up in his book *Cosmology* (Cambridge, UK: Cambridge University Press, 1952). I had the

pleasure of hearing him lecture in Cambridge while I was a graduate student there. He was wonderful.

28 *Paul Dirac*: The English physicist produced the first successful theoretical study of relativistic quantum mechanics (earning him a share of the 1933 Nobel Prize along with Erwin Schrödinger). He made his "large numbers hypothesis" in 1937—pointing to a variety of "coincidences" between ratios of fundamental force scales and universal scales.

29 *remains of a hot big bang*: The universe starts hot, but cools as it expands. Within twenty minutes it cools enough for nucleosynthesis to have produced nuclei-like deuterium, helium, and a little lithium. By about 380,000 years after the Big Bang it's cooled enough for atoms to form when electrons combine with protons and these simple nuclei. This happens because the photons of light streaming through the cosmos no longer have the energy to easily break the electrons free. As a consequence, the photons can whiz on without being snared. As time passes the ever-expanding universe stretches the wavelength of (cools) these photons. By now, 13.8 billion years later, they have been cooled down to microwave wavelengths—and appear to come from all directions in the sky, forming a "background" or sea of radiation.

30 *Carter's own words*: See, for example, Brandon Carter, "Large Number Coincidences and the Anthropic Principle in Cosmology," *Confrontation of Cosmological Theories with Observational Data; Proceedings of the Symposium, Krakow, Poland, September 10–12, 1973*, IAU Symposium No. 63, ed. M.S. Longair (Dordrecht, Netherlands, and Boston: D. Reidel Publishing Company, 1974), 291–98.

30 *veritable gold mine*: I don't mean to imply that a lot of nonsense has been written about the anthropic principle—only some. In a positive light, it is an excellent example of what could be called "selection bias," and it would be foolish to dismiss the ideas without proof. A pretty good overview (in the form of a critique of a book by the physicist Victor Stenger) is Luke Barnes's "The Fine-Tuning of the Universe for Intelligent Life" (2011), available online: http://arxiv.org/abs/1112.4647.

31 *Bernard Carr and Martin Rees*: Their paper was "The Anthropic Principle and the Structure of the Physical World," *Nature* 278 (1979): 605–12.

31 *Rees revisited*: A very nice book by Martin Rees is *Just Six Numbers: The Deep Forces That Shape The Universe* (New York: Basic Books, 2000).

32 *idea of a "multiverse"*: A great deal has been written about the physics that could give rise to multiple universe-like things. Cosmological inflation (an exponential surge in the expansion of the very early universe brought about by a phase change) is one—producing vast numbers of "pocket universes" that are mostly isolated from each other. M-theory,

an extension of string theory, positing that every universe is a multidimensional "brane," or membrane, is another. Possibilities also arise from the "many-worlds" interpretation of quantum mechanics, producing parallel universes with every subatomic event. A great popular account is by Brian Greene, *The Hidden Reality: Parallel Universes and the Deep Laws of the Cosmos* (New York: Alfred A. Knopf, 2011).

32 *changes to our perspective*: After writing this I realized that similar ideas have been talked about before—for example, by the physicist Lee Smolin.

34 *life to fine-tune itself*: Notably, this is a sentiment expressed at various points by the great American paleontologist and evolutionary biologist Stephen J. Gould. It's an interesting perspective. I also wonder: What will we think if we find places in the cosmos that appear perfect for life as we understand it, yet they are sterile?

35 *Fred Hoyle*: Although he suggested the idea in 1953, Hoyle's original paper containing his computation of carbon production in stars was "On Nuclear Reactions Occurring in Very Hot Stars. I. The Synthesis of Elements from Carbon to Nickel," *Astrophysical Journal Supplement* 1 (1954): 121–46.

36 *strongest pieces of evidence*: In later years there has been some discussion about whether Hoyle was really anthropically motivated or just trying to figure out how stars could make carbon. See, for example, the discussion by Helge Kragh, "An Anthropic Myth: Fred Hoyle's Carbon-12 Resonance Level," *Archive for History of Exact Sciences* 64 (2010): 721–51. Kragh's discussion also contains a description of physicist Lee Smolin's general rebuttal of the anthropic use of carbon, not unlike my critique in terms of the "what if" story starting with Galileo.

36 *not quite so fine after all*: This fact has been pointed out by several people, including the physicist Steven Weinberg. Also, a study of the varying energy levels in carbon-12 production in stars indicates that shifts of 60 keV might cause little change in the abundance of carbon produced. See Mario Livio et al., "The Anthropic Significance of the Existence of an Excited State of C-12," *Nature* 340 (1989): 281–84.

36 *something "special" about life*: Saying that life is "special" harks back to ideas of vitalism—the notion that there is some "vital spark" that separates life from nonlife in the universe. Although firmly rejected by mainstream science, these sentiments still manage to sneak back every so often.

2: THE TEN-BILLION-YEAR SPREE

39 *Atacama Desert*: This roughly 600-mile-long region, stretching southward from close to the Chile-Peru border to the north and just to the

west of the full Andean range, includes areas that are considered the driest on Earth (more so than even parts of Antarctica). Indeed, at an altitude of about 10,000 feet there is an area where the aridity and soil chemistry have been compared to conditions on Mars.

39 *La Serena*: La Serena is a city with a population (including immediate surroundings) of a few hundred thousand. Tourism is healthy, thanks to the beaches. It is also the location of administrative offices for the major international astronomical observatories that are sited further inland— those of the United States and Europe.

39 *Elqui Valley*: The valley contains the Elqui River—fed from the Andes and flowing out to the Pacific. Because the area is so dry, the Chileans have now built the Puclaro Dam some forty miles inland to store river water for use during droughts and to try to control flooding during rare storms. The valley is the main Chilean producer of pisco—grape brandy.

40 *Cerro Tololo*: Usually known as CTIO, the observatory is part of the U.S. National Optical Astronomy Observatory, funded by the National Science Foundation. It was established in the early 1960s and is continuously used by both Chilean and U.S.-based scientists.

41 *liquid nitrogen*: Sensitive electronics, including the digital cameras typically used to detect photons and construct images, work better when cooled. Astronomers' fingers, however, do not.

41 *looking for food*: What's on offer at Cerro Tololo is good. The view from the cafeteria is even better—I wish I could sit watching Andean condors swoop by every time I had a meal.

43 *The great Sol*: The diameter of the Sun is generally given for the size of its outermost visible light surface—the so-called photosphere.

45 *the Kuiper Belt*: This region extends from the orbit of Neptune (about thirty times the distance of the Earth from the Sun, or 30 astronomical units, AU) to nearly twice as far again (to about 50 AU). Unlike much of the inner asteroid belt, the majority of objects in the Kuiper Belt are rich in frozen volatiles such as water, methane, and ammonia. All bodies out here and beyond are generally termed trans-Neptunian objects. Although it is named after the Dutch-born astronomer Gerard Kuiper (1905–73), a number of astronomers had speculated on the existence of this region and its contents since Pluto's discovery in 1930.

45 *two and a half thousand*: The energy per unit area from a radiating object decreases as the inverse of its distance squared—a simple geometric effect as light spreads outward in a sphere.

46 *we call the Oort cloud*: Sometimes called the Öpik-Oort cloud, this outlying area of the solar system is named after Dutch astronomer Jan Oort (1900–92), who, among other things, found early evidence in 1932 for the existence of an unseen component of matter in the Milky Way—

what we would today call dark matter. He speculated that long-period comets had to originate in a region very distant from the Sun, but still gravitationally bound to the solar system—the Oort cloud. Dynamics suggest it has an inner zone that is more disk-like and an outer zone that is more spherical.

46 "*false dawn*": This term is thought to have originated with the twelfth-century Persian astronomer, mathematician, and poet Omar Khayyam. In verse 200 of his poetic work *The Rubaiyat*, he wrote:

> *When false dawn streaks the east with cold, gray line,*
> *Pour in your cups the pure blood of the vine;*
> *The truth, they say, tastes bitter in the mouth,*
> *This is a token that the "Truth" is wine.*

48 *produces a thrust*: This is known as the Poynting-Robertson effect, and it's a rather subtle and counterintuitive thing because the mechanism depends on the frame of reference used. One way to conceptualize it is to imagine you're standing in vertical rain. If you walk or run the rain will no longer seem vertical; indeed, it will start to wet your front. An object orbiting the Sun experiences a similar effect with the Sun's light, known as aberration: the illumination seems to be moving slightly *toward* it rather than radially past it. Light carries momentum, and therefore the object (a grain in the zodiacal dust) experiences a drop in its forward or orbital momentum; it gets dragged to a lower orbit. But things are in truth more complicated than this. The object also absorbs radiation, heats up, and emits its own glow of light. For tiny dust grains, the way in which light is absorbed or scattered is also very important, and hinges on the grain composition and actual size. If you're a glutton for punishment, see the excellent but technical journal article by Burns, Lamy, and Soter, which tells all: J. A. Burns et al., "Radiation Forces on Small Particles in the Solar System," *Icarus* 40 (1979): 1–48.

48 *Starting in the 1970s*: See, for example, D. E. Brownlee, D.A. Tomandl, and E. Olszewski, "Interplanetary Dust; A New Source of Extraterrestrial Material for Laboratory Studies," *Proceedings of the Eighth Lunar Science Conference, Houston, Texas, March 14–18, 1977*, Vol. 1 (New York: Pergamon Press, 1977), 149–60.

48 *benign dispersal of comets*: See, for example, D. Nesvorný et al., "Dynamical Model for the Zodiacal Cloud and Sporadic Meteors," *The Astrophysical Journal* 743 (44): 129–44.

49 *Recently the Hubble Space Telescope*: Other observatories also collaborated; see, for example, D. Jewitt et al., "*Hubble Space Telescope* Observations of Main-Belt Comet (596) Scheila," *The Astrophysical*

Journal Letters 733 (2011): L4–L8; and J. Kim et al., "Multiband Optical Observation of the P/2010 A2 Dust Tail," *The Astrophysical Journal Letters* 746 (2012): L11–L15.

50 *one-in-a-billion*: When energy is converted into subatomic particles, it produces pairs—one of matter, one of antimatter—that will annihilate each other and revert to electromagnetic energy if they combine. Yet we seem to exist in a universe dominated by matter. Thus, in the very early universe, barely a millionth of a second old, there must have been a slight asymmetry between matter and antimatter so that as the universe cooled it was left with unpaired particles of matter. A billion particles of antimatter existed for every billion+1 particles of matter. Why? Good question. We don't yet know, although particle physics experiments using large colliders seem to be homing in on the answer.

50 *windblown sand*: This is actually not too far from the truth. Recent studies of circumstellar dust indicate that some is highly resilient, made of silicates (magnesium silicate, for example), and blown outward from stars by radiation pressure.

50 *called the Trifid Nebula*: See, for example, the article by J. J. Hester et al., "The Cradle of the Solar System," *Science* 304 (2004): 1116–17.

52 *egg-like patches*: The structure of these patches really is quite egg-like, and astronomers often use the name "propylid" (a disk of dense gas and dust surrounding a young star)—derived from the fact that these are, or are precursors to, protoplanetary disks.

52 *Fluffy and sticky*: The study of protoplanetary and interplanetary dust/ particle aggregates suggests they are quite loose, not unlike the dust bunnies some of us find under our beds.

52 *the "snow line"*: In a vacuum water ice sublimates (evaporates) very rapidly at temperatures above about 150–170 Kelvin; therefore, the snow line occurs at distances from the system center where objects can cool below these temperatures.

53 *marvelous mash of chemistry*: We know this because we can make telescopic observations that examine the radiation from a protoplanetary or circumstellar disk and analyze the discrete features in the spectra of this radiation that are fingerprints of known atoms and molecules.

54 *giant sputtering engine*: One of the stages between an early protostellar system and a hydrogen-fusing star (a so-called zero-age main sequence star) is known as a T-Tauri star (after the prototype). These objects appear to be slowly contracting and heating up due to gravitational forces, prone to sporadic outbursts of radiation, and on their way to finally settling down to steady fusion.

54 *seven-hour period*: See, for example, D. A. Clarke, "Astronomy: A Truly Embryonic Star," *Nature* 492 (2012): 52–53.

55 *the Allende and the Murchison*: Anecdotal (but very believable) reports suggest that since this was 1969—at the cusp of the Apollo program's breakthrough Moon landing—public and scientific interest in these events was intense, and perhaps helped with the rapid collection of many fragments from both meteorites.

56 *older than the Earth itself*: See, for example, A. Bouvier, M. Wadhwa, "The Age of the Solar System Redefined by the Oldest Pb–Pb Age of a Meteoritic Inclusion," *Nature Geoscience* 3 (2010): 637–41.

56 *isotope of magnesium*: A very good popular account of the isotopic clues in meteorites, and much more, is Jacob Berkowitz, *The Stardust Revolution: The New Story of Our Origin in the Stars* (New York: Prometheus Books, 2012).

56 *must have exploded*: For many details, including the effects on our nascent solar system, see the excellent review by N. Dauphas and M. Chaussidon, "A Perspective from Extinct Radionuclides on a Young Stellar Object: The Sun and Its Accretion Disk," *Annual Review of Earth and Planetary Sciences* 39 (2011): 351–86. See also Y. Lin et al., "Short-Lived Chlorine-36 in a Ca- and Al-Rich Inclusion from the Ningqiang Carbonaceous Chondrite," *Proceedings of the National Academies of Sciences of the United States [PNAS]* 102 (2005): 1306–11.

58 *stellar Eden*: A great (technical) review of the Sun's birth environment is F. Adams, "The Birth Environment of the Solar System," *Annual Review of Astronomy and Astrophysics* 48 (2010): 47–85.

58 *Messier 67*: We don't yet know the final conclusion on whether this stellar cluster is the Sun's birthplace or not. There are certainly very close Sun "analogues" in this system (stars of similar composition), but (see below) the motions and orbits of objects may not allow it.

59 *recent computer simulations*: See, for example, B. Pichardo et al., "The Sun Was Not Born in M67," *The Astronomical Journal* 143 (2012): 73–83.

60 *metallic liquid state*: Yes, under sufficient pressure hydrogen behaves like a metal. Inside Jupiter is some fifty times the mass of the Earth in the form of metallic hydrogen.

62 *Earth-like climates*: I don't much like the term "Earth-like" (as you'll see later), but it's a very convenient catchall. In this case the emphasis is on "like," because while a planet such as Mars seems to have harbored liquid water on its surface at certain times, it may have always had a climate more akin to a nasty cold desert than to anything tropical.

62 *sailing away to space*: The technical term is Jeans escape, where the velocity of an atom or molecule equals the escape velocity from the gravity well of a planet at that altitude. There are other loss mechanisms, too, including "sputtering," where energetic particles of solar wind (such

as protons) slam into the atmospheric atoms or molecules and literally knock them off into space.

63 *encase the planet in ice*: One example of a Snowball Earth is suggested by the rock record from around 650 million years ago.

66 *planets have survived*: Exactly which ones is debated. While the general consensus is that Mercury and Venus will be engulfed by our dying star, it is not clear whether or not Earth will survive. I've chosen to be optimistic here. A less optimistic view is K. Rybicki, C. Denis, "On the Final Destiny of the Earth and the Solar System," *Icarus* 151 (2001): 130–37.

3: NEIGHBORS

69 *tales of* One Thousand and One Nights: You can of course read these for yourself. A really excellent resource is Robert Irwin's *The Arabian Nights: A Companion* (New York: Viking Adult, 1994; rev. ed., London: Tauris Parke Paperbacks, 2004).

70 *allegorical Narnia:* What's interesting about Narnia (*The Lion, the Witch and the Wardrobe*) and *Star Wars* is how both have a "savior" story. While C. S. Lewis's was clearly a Christian allegory, George Lucas's was a more worldly amalgam of all the best parts of fairy tales and spiritual fables. And both take place "somewhere else," where earthbound rules seldom apply.

70 *Finding planets around other stars*: There are many good accounts of the search for exoplanets. A few are: Alan Boss, *The Crowded Universe: The Search for Living Planets* (New York: Basic Books, 2009); Ray Jayawardhana, *Strange New Worlds: The Search for Alien Planets and Life beyond Our Solar System* (Princeton: Princeton University Press, 2011); Lee Billings, *Five Billion Years of Solitude* (New York: Current/Penguin, 2013).

71 *Doppler effect*: Named after the nineteenth-century Austrian physicist Christian Doppler, this is an effect on the frequency of a wave due to relative motion. A common example is the increase in pitch (frequency) of a siren on a police car or ambulance as it comes toward you, in effect compressing its sound waves, and its decrease in pitch as it moves away from you, stretching its sound waves out. The "red shift" of stars and galaxies moving away from us is the equivalent as it applies to electromagnetic radiation or light, but the fact that light moves at, well, the speed of light requires some time-bending adjustments, which are fully described by the equations of the relativistic Doppler effect.

72 *eclipsing their parent stars*: This is known as the transit method: the planets transit, or pass in front of, their parent stars, blocking a small amount of light. It has become a major technique for spotting other

worlds, exemplified by the COROT and Kepler space telescopes. Careful analysis of variations in the timing of transits can also reveal the presence of other, non-transiting planets in a system that are tugging at the ones we can spot.

72–73 *gravitational lensing*: The presence of planets can produce some weird and wonderful, and complex, variations in the way the light from a background star is seen. However, the rate at which, from our point of view, stars with planets line up with more-distant stars (also perhaps with planets) to produce the lensing effect is low. So gravitational lensing searches require a lot of patience and careful monitoring of many, many stars. Nonetheless, this technique is unique in its sensitivity to planets at a wide range of orbital distances from their stars and provides vital statistics on planet abundances.

73 *bold and persistent*: Among the names that are sometimes forgotten (although many have also become rightfully famous, particularly Michel Mayor, Didier Queloz, Geoff Marcy, and R. Paul Butler) are the Canadians Gordon Walker and Bruce Campbell, who pioneered the modern techniques of Doppler detection.

74 *Titius-Bode law*: This empirical rule predicting the spacing of planetary orbits is named for German astronomers Johann Titius (1729–96) and Johann Bode (1747–1826)—the latter largely responsible for promoting the idea. The "law" doesn't really work for Neptune, with a 30 percent difference between its prediction for orbital semimajor axis and the actual value. Nonetheless, the Titius-Bode Law is still invoked for certain exoplanetary systems as a convenient rule of thumb—since planets tend to have their orbits spaced uniformly in the logarithm of radius (distance from the star) due to the general nature of planet formation. I'm not convinced that it's a good thing that we hang on to the idea, though, since we have an incomplete physical understanding of these processes.

75–76 *Arecibo Observatory*: The primary research facility of the National Astronomy and Ionosphere Center (NAIC), Arecibo was constructed in the early 1960s and completed in 1963. It has played a role in many major scientific discoveries since then, including millisecond pulsars, binary pulsars, and radar imaging of the surface of Venus.

77 *Aleksander Wolszczan and Dale Frail*: The discovery is reported in A. Wolszczan, D. Frail, "A Planetary System around the Millisecond Pulsar PSR1257 + 12," *Nature* 355 (1992): 145–47.

77 *at least* three *planet-sized*: Although there have been claims of a fourth body, these results seem to be in doubt; see, for example, A. Wolszczan, "Discovery of Pulsar Planets," *New Astronomy Reviews* 56 (2012): 2–8.

79 *in 1995, astronomers*: The name of the star is 51 Pegasi, and the two key publications/announcements are M. Mayor, D. Queloz, "A Jupiter-Mass Companion to a Solar-Type Star," *Nature* 378 (1995): 355–59, and the

confirmation: M. Mayor, D. Queloz, G. Marcy, P. Butler et al., "51 Pegasi," *International Astronomical Union Circular* 6251 (1995): 1.

80 *Peter Goldreich and Scott Tremaine*: Their paper on orbital migration is P. Goldreich, S. Tremaine, "Disk-Satellite Interactions," *The Astrophysical Journal* 241 (1980): 425–41.

81 *barely twenty-four Earth hours*: A great resource for exploring the amazing diversity of exoplanets is the constantly updated online catalog maintained at http://exoplanet.eu/catalog/ and originated by Jean Schneider at the Observatoire de Paris.

81 *fearsome climates*: See, for example, I.A.G. Snellen et al., "The Orbital Motion, Absolute Mass and High-Altitude Winds of Exoplanet HD209458b," *Nature* 465 (2010): 1049–51.

81 *all manner of chemistry*: Planetary atmospheres are complicated things to understand. A good reference for some of the work on "hot Jupiter" properties is A. Burrows, J. Budaj, I. Hubeny, "Theoretical Spectra and Light Curves of Close-in Extrasolar Giant Planets and Comparison with Data," *The Astrophysical Journal* 678 (2008): 1436–57.

82 *against the spin sense*: This strange retrograde motion was first seen in the system WASP-17b, as described by D. Anderson et al., "WASP-17b: An Ultra-Low Density Planet in a Probable Retrograde Orbit," *The Astrophysical Journal* 709 (2010): 159–67.

83 *Pitch-black planets*: See D. M. Kipping, D. S. Spiegel, "Detection of Visible Light from the Darkest World," *Monthly Notices of the Royal Astronomical Society* 417 (2011): L88–L92.

83 *"Icarus worlds"*: For example, the gas giant orbiting star HD 8606 (190 light-years from Earth) has an orbital period of 111 Earth days, but an ellipticity of 0.93. This means its closest approach to the star comes within 0.03 astronomical units, while its farthest distance is 0.88—a factor of thirty. During its closest pass the atmospheric temperature is thought to rise by a factor of two in just six hours.

84 *exhaust from a cosmic smelter*: See, for example, S. Rappaport et al., "Possible Disintegrating Short-Period Super-Mercury Orbiting KIC 12557548," *The Astrophysical Journal* 752 (2012): 1.

85 *nine major planets*: It'd be fair to say that we're not yet sure we've found systems precisely like this, because interpreting the data is very tricky. Nonetheless, I base this hypothetical set of planets on a real claim by M. Tuomi, "Evidence for Nine Planets in the HD 10180 System," *Astronomy and Astrophysics* 543 (2012), no. A52:1–12.

87 *are the super-Earths*: See for example the review by N. Haghighipour, "The Formation and Dynamics of Super-Earth Planets," *Annual Review of Earth and Planetary Sciences* 41 (2013): 469–95.

88 *may be hundreds of billions*: See, for example, X. Bonfils et al., "The

HARPS Search for Southern Extra-Solar Planets. XXXI. The M-dwarf Sample," *Astronomy and Astrophysics* 549, no. A109 (2013): 1–75.

89 *such stars are so dim*: And none are close enough to be seen with the unaided human eye.

89 *more than a* trillion *years*: A good reference for the theory behind this statement is G. Laughlin, P. Bodenheimer, F. C. Adams, "The End of the Main Sequence," *The Astrophysical Journal* 482 (1997): 420–32.

90 *evidence for a remarkable number*: The majority of this evidence comes from gravitational microlensing surveys. See T. Sumi et al. and A. Udalski et al. (the Microlensing Observations in Astrophysics [MOA] and Optical Gravitational Lensing Experiment [OGLE] collaborations), "Unbound or Distant Planetary Mass Population Detected by Gravitational Microlensing," *Nature* 473 (2011): 349–52.

90 *Twin suns, sometimes even more*: Many known exoplanets orbit one star which in turn has one or even two companion stars in more-distant orbits. For example, the system GJ667 has three stars (A, B, C), and star C has five confirmed exoplanets orbiting it. The best-confirmed case of a planet orbiting two stars at once is Kepler-16, sometimes known as the "Tatooine" system in honor of a fictional place depicted in the movie *Star Wars*.

91 *ocean planets*: See A. Léger et al., "A New Family of Planets? 'Ocean-Planets,'" *Icarus* 169 (2004): 499–504.

92 *My colleagues and I*: Collaborations I was a part of produced a series of papers from 2008 to 2010 studying the rudiments of planetary climate variations. The first of these is D.S. Spiegel, K. Menou, and C.A. Scharf, "Habitable Climates," *The Astrophysical Journal* 681 (2008): 1609–23.

93 *the term "Earth-like"*: I published a very public takedown of this notion online at *Scientific American*: "Should We Expect Other Earth-Like Planets At All?," December 26, 2012, http://blogs.scientificamerican .com/life-unbounded/2012/12/26/should-we-expect-other-earth-like -planets-at-all/.

94 *a galaxy-wide population*: Two good references on extrapolations to make these statements about the total abundance of planets in the Milky Way: C. D. Dressing, D. Charbonneau, "The Occurrence Rate of Small Planets around Small Stars," *The Astrophysical Journal* 767 (2013): 95– 114; and E. A. Petigura, G. W. Marcy, and A. W. Howard, "A Plateau in the Planet Population below Twice the Size of Earth," *The Astrophysical Journal* 770 (2013): 69–89.

95 *important yet subtle thing*: I have written more about this in *Aeon Magazine*: C. Scharf, "Are We Alone?," June 20, 2013, http://aeon.co/maga zine/nature-and-cosmos/the-real-meaning-of-the-exoplanet-revolution/.

4: A GRAND ILLUSION

97 *Henri Poincaré*: Poincaré (1854–1912) wasn't just a mathematician; he excelled at pretty much everything he put his mind to, including physics and engineering. Most accounts indicate that he was prone to working quickly and with relatively little concern for making corrections or alterations to his work.

97 *journal* Acta Mathematica: The journal is still going strong, published by the Institut Mittag-Leffler, a research institute of the Royal Swedish Academy of Sciences.

98 *"n-body problem"*: This famous problem in mathematical physics can be found throughout the research literature. There are a number of (highly contrived) examples of exact solutions for very special cases: see, for example, Cristopher Moore, "Braids in Classical Dynamics," *Physical Review Letters* 70 (1993): 3675–79, and some wonderful animations of these at http://tuvalu.santafe.edu/~moore/gallery.html.

100 *King Oscar's competition*: For the time line and story of Poincaré's efforts, an excellent brief essay with extensive references is by Q. Wang, "On the Homoclinic Tangles of Henri Poincaré," http://math.arizona.edu/~dwang/history/Kings-problem.pdf.

100 *cost anyone so much*: The prize money was 2,500 kronor, compared to the 3,500 kronor needed to reprint the copies of *Acta Mathematica*. For comparison, at the time a typical annual salary for a Swedish academic was about 7,000 kronor.

101 *In the 1990s*: A great essay on this later history of the n-body problem is by F. Diacu, "The Solution of the n-body Problem," *The Mathematical Intelligencer* 18 (1995): 6670.

102 *called nonlinearity*: If you want to learn more about the many, many aspects of chaos and nonlinearity, the book by James Gleick is still a superb read: *Chaos: Making a New Science* (New York: Viking Penguin, 1987; rev. ed., Penguin Books, 2008).

103 *Jacques Laskar*: The paper on this work was J. Laskar, "A Numerical Experiment on the Chaotic Behaviour of the Solar System," *Nature* 338 (1989): 237–38.

103 *Gerald Sussman and Jack Wisdom*: The paper on this work was G. J. Sussman, J. Wisdom, "Chaotic Evolution of the Solar System," *Science* 257 (1992): 56–62.

104 *exponential divergence*: This property is quantified by the "Lyapunov exponent," a mathematical quantity that characterizes the rate at which infinitesimally different trajectories (e.g., orbits) in a dynamic system move apart from each other—in other words, how rapidly the system becomes unpredictable. It is named after the Russian scientist Aleksandr Lyapunov (1857–1918).

106 *harvest these futures*: More recent work includes the study of the effect of Einstein's general relativity on the orbital evolution of our solar system—which provides corrections to Newton's simple force laws for gravity. See, for example, G. Laughlin, "Planetary Science: The Solar System's Extended Shelf Life," *Nature* 459 (2009): 781–82. And J. Laskar, M. Gastineau, "Existence of Collisional Trajectories of Mercury, Mars and Venus with the Earth," *Nature* 459 (2009): 817–19.

106 *Konstanin Batygin and Greg Laughlin*: Their paper is "On The Dynamical Stability of the Solar System," *The Astrophysical Journal* 683 (2008): 1207–16.

110 *only about twenty-one hours*: See, for example, G. E. Williams, "Geological Constraints on the Precambrian History of Earth's Rotation and the Moon's Orbit," *Reviews of Geophysics* 38 (2000): 37–59.

111 *crafted computer codes*: There are many of these, each with its own approach and often with its own specialized application (whether it be planets or galaxies). They have names like "Mercury," "SWIFT," and "Hermit."

113 *period of youthful chaos*: Many research articles have dealt with this topic. See for example F.C. Adams, G. Laughlin, "Migration and Dynamical Relaxation in Crowded Systems of Giant Planets," *Icarus* 163 (2003): 290–306; M. Juric, S. Tremaine, "Dynamical Origin of Extrasolar Planet Eccentricity Distribution," *The Astrophysical Journal* 686 (2008): 603–20.

114 *A leading theory*: This theory is known as the "Nice model" after the Observatoire de la Côte d'Azur, Nice, France, where it was developed. See, for example, K. Tsiganis et al., "Origin of the Orbital Architecture of the Giant Planets of the Solar System," *Nature* 435 (2005): 459–61.

115 *David Nesvorný*: His paper discussing a fifth giant planet is "Young Solar System's Fifth Giant Planet?," *The Astrophysical Journal Letters* 742 (2011): L22–L27.

116 *more than 60 percent*: See, for example, A. Cassan et al., "One or More Bound Planets per Milky Way Star from Microlensing Observations," *Nature* 481 (2012): 167–69.

117 *"habitable zones"*: This is a topic with an enormous literature, many fascinating ideas, and almost no consensus over the details of how to estimate whether a given planet is capable of supporting life or not. That being said, a good place to start would be to read James Kasting's insightful book *How to Find a Habitable Planet* (Princeton: Princeton University Press, 2010).

118 *30 percent fainter*: This has become known as the "faint young Sun problem," and it's still not solved, despite a constant stream of works claiming to have an answer. You can read a review by G. Feulner, "The

Faint Young Sun Problem," *Reviews of Geophysics* 50 (2012): RG2006. My personal hunch: better (3-D) climate models may solve the problem by more accurately describing planetary climate. My unsubstantiated pet theory: perhaps Earth's orbit wasn't quite what we think it was.

119 *known as Theia*: In the so-called giant impact theory of the Moon's formation, a roughly Mars-size object, a protoplanet called Theia ("goddess"), occupied the same orbital range as the young Earth—perhaps a "horseshoe" orbit back and forth around one of the stable points (Lagrange points) leading or trailing Earth's own orbit. Eventually, its orbital path brought it into collision with the Earth. Although this is the leading theory at the moment, there are signs that it may be an incomplete picture of what happened. See, for example, this short review by D. Clery, "Impact Theory Gets Whacked," *Science* 342 (2013): 183–85.

121 *Hal Levison and his colleagues*: Their study is reported in H. F. Levison et al., "Capture of the Sun's Oort Cloud from Stars in Its Birth Cluster," *Science* 329 (2010): 187–90.

5: SUGAR AND SPICE

126 *a lesson in abject humility*: Archaea, like bacteria, are prokaryotes, single-celled organisms with no cell nucleus or other organelles. In 1977 for the first time certain species were recognized as a distinct type of prokaryote and classified in their own kingdom, separate from the bacteria, by Carl R. Woese and George E. Fox, on the basis of genetic analyses reported in "Phylogenetic Structure of the Prokaryotic Domain: The Primary Kingdoms," *PNAS* 74 (1977): 5088–90.

127 *the million trillion trillion*: Not surprisingly, these estimates vary. The number I quote is based on an influential paper by William B. (Brad) Whitman, "Prokaryotes: The Unseen Majority," *PNAS* 95 (1998): 6578–83. It involves a number of educated extrapolations from measured populations and environments.

128 *my previously narrow ideas*: The paper by Paul Falkowski, Tom Fenchel, and Edward Delong is "The Microbial Engines That Drive Earth's Biogeochemical Cycles," *Science* 320 (2008): 1034–39.

128 *"multimeric protein complexes"*: Many molecular machines consist of these proteins containing two or more, same or different, polypeptide chains. A polypeptide is basically a chain of amino acids, held together by covalent bonds—which are the result of a sharing of electrons between atoms. Phew, chemistry is complicated.

131 *combinations of molecular "fuel"*: A very nice review of life's energy budgets and the fuel/burning way of looking at things is by K. H. Nealson and P. G. Conrad, "Life: Past, Present, and Future," *Philosophical*

Transactions of the Royal Society B: Biological Sciences 354 (1999): 1923–39.

131 *methanogenesis*: Although it may sound straightforward here, the formation of methane by microbes, like all metabolic processes, involves a bewildering array of enzymes and reactions, and not always the same ones. In fact, there are three major metabolic routes that produce methane: the reduction of carbon dioxide (which is emphasized here), the fermentation of acetate, and the dismutation (simultaneous oxidation and reduction to make two products) of methanol or methylamines. Each involves multiple chemical steps.

131 *spread around the entire planet*: The examples are numerous. A recent, and particularly fascinating, one is the apparent coupling together of chemical reactions (redox reactions) between widely separated (12 millimeters, a vast gulf for bacteria) layers of marine sediment. A possible mechanism to link these physical layers is actually electrical—bacteria may really be controlling the flow of charged particles through the planet. L. P. Nielsen et al., "Electric Currents Couple Spatially Separated Biogeochemical Processes in Marine Sediment," *Nature* 463 (2010): 1071–74.

133 *photosynthesis was a metabolic tool*: Types of cyanobacteria (blue-green bacteria) were using sunlight to make their own food more than 3 billion years ago. These oxygen-producing organisms are still ubiquitous on Earth.

133 *deep-sea hydrothermal vent systems*: See, for example, N. Lane, W. F. Martin, "The Origin of Membrane Bioenergetics," *Cell* 151 (2012): 1406–16.

133 *horizontal gene transfer*: Bacteria, for example, can exchange small subsets of their genetic material in the form of "plasmids." These plasmids often exist as circular strands of DNA (separate from chromosomal DNA) in the cell, with genetic codes anywhere between a thousand and a million base pairs (letters) long. Why does nature do this? One advantage for microbes is the ability to share DNA that codes for resistance to threats like antibiotics. In effect, the dispersal of plasmids increases the chances of survival for entire populations, not just for an individual who happens to have the right mutation.

135 *"Snowball Earth" episode*: This is still a somewhat controversial idea. There is evidence from the rock record that sometime between 650 and 750 million years ago there may have been a period of globally low temperatures, possibly cold enough to bring permanent ice cover even to the lowest latitudes. The extent to which Earth froze up, why, and how it recovered are all subjects of debate. A pro-snowball paper is, for example, P. F. Hoffman et al., "A Neoproterozoic Snowball Earth," *Science*

281 (1998):1342–46. It is true that planets with surface water are susceptible to a positive-feedback process in which frozen water reflects more solar energy than liquid water—thereby further lowering the surface temperature. Snowball states may be common among exoplanets.

138 *H_3^+ is remarkable*: See, for example, the discussion by B. J. McCall and T. Oka, "H_3^+—an Ion with Many Talents," *Science* 287 (2000): 1941–42.

138 *Methanol, ethanol, and acetylene*: See D. F. Strobel, "Molecular Hydrogen in Titan's Atmosphere: Implications of the Measured Tropospheric and Thermospheric Mole Fractions," *Icarus* 208 (2010): 878–86 (and references therein).

139 *chemistry of life on Earth*: Indeed, some work on the more abstract systems structure of metabolism and carbon chemistry suggests that carbon-based metabolism was a near certainty, an "attractor" in the space of possibilities. See R. Braakman and E. Smith, "The Compositional and Evolutionary Logic of Metabolism," *Physical Biology* 10 (2012): 011001.

141 *the Saturnian moon Titan*: Measurements of a downward flux of molecular hydrogen in Titan's atmosphere have led to some renewed discussion of life on the moon. See Strobel, "Molecular Hydrogen."

142 *a biological census*: Using the tools of "metagenomics," environmental samples are processed in order to measure the genetic diversity present in certain critical genes that all life utilizes. For example, the 16S ribosomal RNA gene sequence consists of 1,542 nucleic acid "letters," and is what's termed "highly conserved"—meaning that random mutations tend to cause problems and are quickly discarded through natural selection—therefore any different versions typically correspond to different species of organisms. Measuring the diversity of this sequence in a sample yields an estimate of the number of distinct species of bacteria or archaea that are present.

142 *inside our lungs*: See, for example, the review by J. M. Beck, V. B. Young, and G. B. Huffnagle, "The Microbiome of the Lung," *Translational Research* 160 (2012): 258–66.

143 *genetic census of the human stomach and gut*: There are various excellent sources of information on this incredible field of exploration. A great nontechnical discussion is by J. Ackerman, "The Ultimate Social Network," *Scientific American* 306 (2012): 36–43. Although new work is constantly being announced on the human microbiome, the 2010 research on gut microbes was part of the MetaHIT project (Metagenomics of the Human Intestinal Tract), and reported by J. Qin et al. in "A Human Gut Microbial Gene Catalogue Established by Metagenomic Sequencing," *Nature* 464 (2010): 59–65.

144 *three major types*: The enterotypes were identified by analyzing the metagenomic data from (this is lovely) human fecal matter. The work is described by M. Arumugam et al. in "Enterotypes of the Human Gut Microbiome," *Nature* 473 (2011): 174–80.

145 *"microbial soul"*: It's very early days for research into this possibility, but it does seem that there are connections, sometimes termed the "microbiome-gut-brain-axis." A good overview is by V. O. Ezenwa et al., "Animal Behavior and the Microbiome," *Science* 338 (2012): 198–99.

145 *our own evolution*: Indeed, this has led to some scientists using the term "hologenome"—the sum of human plus microbiome genes as the quantity to study in terms of evolution and natural selection (true for any multicellular organism). There is research that seems to be supporting some of these ideas; see, for example, R. M. Brucker, S. R. Bordenstein, "The Hologenomic Basis of Speciation: Gut Bacteria Cause Hybrid Lethality in the Genus *Nasonia*," *Science* 341 (2013): 667–69.

148 *creatures like ants*: Probably the definitive work is still that by B. Hölldobler and E. O. Wilson, *The Ants* (Cambridge, MA: Belknap Press of Harvard University Press, 1990).

149 *materials such as coconut shells*: The first academic report of octopi gathering "tools" for later use was made by J. K. Finn, T. Tregenza, and M. D. Norman, "Defensive Tool Use in a Coconut-Carrying Octopus," *Current Biology* 19 (2009): R1069–70. They observed veined octopi collecting, stacking, and transporting (awkwardly, using a special "stilt-walking" maneuver) coconut shells—seemingly hoarding them for use as shelter. Finn recounts that the sight was comical: "I have never laughed so hard underwater." www.eurekalert.org/pub_releases/2009 -12/cp-tui120909.php.

149 *123,000 and 195,000 years ago*: Precise dates are not known. This time frame is established from evidence of a glacial stage known as Marine Isotope Stage 6 (MIS6) and from studies of genetic diversity in humans. Other potential bottlenecks in human population have also been proposed—for example, some 70,000 years ago, and even back 1.2 million years ago. However, I think it's safe to say that not everyone agrees that such declines in population happened. A good overview is G. Hewitt, "The Genetic Legacy of the Quaternary Ice Ages," *Nature* 405 (2000): 907–13.

150 *living off rich pickings*: As with all such statements, not everyone agrees with this interpretation. Judge for yourself; a good reference is C. W. Marean et al., "Early Human Use of Marine Resources and Pigment in South Africa During the Middle Pleistocene," *Nature* 449 (2007): 905–908.

150 *drifted to extinction about 28,000 years ago*: This is a rough figure. There seems to be varying opinion on the demise of Neanderthals—especially in terms of geographical location.

150 *genetic sequence of Neanderthal remains*: See R. E. Green et al., "A Draft Sequence of the Neanderthal Genome," *Science* 328 (2010): 710–22.

151 *sequences of DNA that differ most*: An excellent review is the popular piece by K. S. Pollard, "What Makes Us Different?," *Scientific American* 22 (2012): 30–49. I have drawn heavily on this.

152 *being intelligent has been part of*: I think this has to be true at a certain level; however, it has also been pointed out that if human-style intelligence is so great, why does it seem to have arisen only once on Earth in 4 billion years? I'm not sure how good this argument is, though. For example, flowering plants are incredibly successful in evolutionary terms but have only arisen once, some 130 million years ago, and this may well be related to populations of organisms such as insects. As always, a multitude of factors contribute to the success or failure of particular biological strategies.

6: HUNTERS OF THE COSMIC PLAIN

155 *the most ancient cave paintings*: I'm thinking in particular of the extraordinary Chauvet-Pont-d'Arc Cave in the Ardèche department of southern France, containing some of the most stunning paintings of hundreds of animals, dating as far back as 30,000 to 32,000 years ago. A beautiful visual study is Werner Herzog's documentary *Cave of Forgotten Dreams* (2010).

156 *endless cosmic repetition and rebirth*: The idea of a cyclical cosmos prevails—for example, in Hindu philosophy and in Buddhism.

156 *never been any data*: At the time of writing we have absolutely no data about life elsewhere. Of course the absence of data is itself interesting, and has certainly been used in constructing theories about the nature of spacefaring life in the universe, and why it hasn't shown up yet (it hasn't, despite a lot of wishful thinking). I discuss this puzzle in the final chapter.

157 *William Herschel*: The German-born British scientist was a tremendous astronomer, optical engineer, and even composer. The statements he made about life on the Moon or Sun are drawn in part from Iwan Rhys Morus's book *When Physics Became King* (Chicago: The University of Chicago Press, 2005). Herschel's own papers are also useful: for example, W. Herschel, "On the Nature and Construction of the Sun and Fixed Stars," *Philosophical Transactions of the Royal Society of London* 85 (1795): 46–72, and some of his thoughts on the Moon in W. Herschel,

"Astronomical Observations Relating to the Mountains of the Moon," *Philosophical Transactions* 70 (1780): 507–26.

158 *Thomas Dick*: A useful reference to both Dick's and Herschel's ideas about pluralism is Michael J. Crowe's *The Extraterrestrial Life Debate, 1750–1900: The Idea of a Plurality of Worlds from Kant to Lowell* (Cambridge, UK: Cambridge University Press, 1986).

160 *more than a billion trillion stars*: Estimating the total number of stars in the observable universe is not an exact science. The number quoted here is actually rather conservative at 10^{21}; some studies suggest it could be three hundred times larger. This is what's arrived at by extrapolating from studies such as P. G. van Dokkum and C. Conroy, "A Substantial Population of Low-Mass Stars in Luminous Elliptical Galaxies," *Nature* 468 (2010): 940–42.

161 *Thomas Bayes*: A lot has been written about Bayes, especially since the growth in the use of Bayesian statistics in the past few decades. One source I've made use of here has been D. R. Bellhouse's essay "The Reverend Thomas Bayes, FRS: A Biography to Celebrate the Tercentenary of his Birth," *Statistical Science* 19 (2009): 3–43. A more popular account is by Sharon Bertsch McGrayne, *The Theory That Would Not Die: How Bayes' Rule Cracked the Enigma Code, Hunted Down Russian Submarines, and Emerged Triumphant from Two Centuries of Controversy* (New Haven: Yale University Press, 2011).

162 *Richard Price*: Price deserves a lot more credit than he usually gets, for being instrumental in getting Bayes's ideas into publishable form and for seeing them in a philosophical light.

162 *"Bayes' theorem"*: In simple form it is: $P(A|B) = \frac{P(B|A)P(A)}{P(B)}$. The probability of A given B is the product of the probability of B given A and the probability of A, divided by the probability of B, where (for example) A may be the hypothesis or model, and B is the data.

162 *sketched out as a note*: The example Price used was of a newborn baby watching the Sun rise and set. I prefer chickens.

164 *Cheshire cats*: What really inspired Lewis Carroll to create these iconic beasts (or one of them, at least) is not known. Theories range from heraldic pictures of lions, to church gargoyles, to folklore about happy, milk-fed cats from Cheshire in England.

168 *such sloppiness*: This "debate" fell into two camps: one was that of the so-called frequentists, the other the Bayesians. A frequentist interprets things on the basis of the outcome of measurements, and typically assumes that there are fixed underlying parameters that one cannot assign probabilities to. For example, if an experiment yields a certain result in 95 out of 100 cases, frequentists might state that 95 percent of any further experiments will also yield that result—they don't assign probabilities.

170 *David Spiegel and Edwin Turner*: Their paper is "Bayesian Analysis of the Astrobiological Implications of Life's Early Emergence on Earth," *PNAS* 109 (2012): 395–400.

170 *life appeared quite early*: The most generally accepted evidence of early life on Earth comes from stromatolites—layered rocky deposits made by microbial colonies. Such structures are still being formed in a handful of special locations, such as Shark Bay in Australia and the Exuma Cays in the Bahamas. The oldest that seem to be plausibly biological in origin are some 3.45 billion years old. Claims also exist for weblike imprints from microbial colonies in Australia that may be 3.49 billion years old. There have also been claims of 3.8-billion-year-old microbial deposits, but these are more controversial. Part of the difficulty in establishing the earliest life is simply that there are few places where such very ancient rocks are accessible.

174 *on its way to Pluto*: This is NASA's New Horizons mission, launched in 2006 and due to fly past Pluto and its moons in 2015 at about 14 kilometers per second, before heading on to other targets.

174 *tens of thousands of years*: This is not true for all probes. *Pioneer 10* may take more than 60 million years to pass moderately close to the star Aldebaran (68 light-years away). *Pioneer 11* should pass within about 1.7 light-years of a low-mass star in about 40,000 years. *Voyager 1* will also pass within a couple of light-years of another low-mass star in about 40,000 years, and *Voyager 2* will pass a few light-years from the star Sirius in some 296,000 years.

174 *Jovian moon Europa*: Visual inspection of structures on this moon's frozen water surface, and the detection of sulfate salts on the surface that must have started as chloride salts, together with measurements of induced magnetic fields, all point toward a substantial subsurface ocean on Europa. It is likely beneath a solid crust several tens of kilometers thick, but may occasionally leak out through tectonic-like processes. What helps keep Europa's interior warm is probably a mix of radioactive heating from a rocky core and the frictional heating from gravitational tides as Europa is stretched and relaxed due to its elliptical orbit around massive Jupiter (an orbit produced by interaction with the other Galilean satellites).

175 *LUCA, a single species*: For a good but slightly older overview see D. Penny and A. Poole, "The Nature of the Last Universal Common Ancestor," *Current Opinion in Genetics and Development* 9 (1999): 672–77. The Bayesian analyses that support LUCA are reported by D. L. Theobald, "A Formal Test of the Theory of Universal Common Ancestry," *Nature* 465 (2010): 219–22. There is also an excellent discussion of that work by M. Steel and D. Penny, "Origins of Life: Common Ancestry Put to the Test," *Nature* 465 (2010): 168–69.

176 *"RNA world"*: A great deal has been written about this idea, which goes back to the 1960s under various guises. The first use of the term "RNA world" was by Walter Gilbert, "Origin of Life: The RNA World," *Nature* 319 (1986): 618.

176 *claims to have found fossil cells*: (See the earlier note on microbe-produced structures in rocks.) Paleontologists have recently claimed to find 3.4-billion-year-old fossils of sulfur-using bacterial cells, and (separately) 3.49-billion-year-old microbially produced weblike patterns in rocks, both from sites in the Pilbara in Western Australia.

177 *giant viruses:* These beasties really are overturning a lot of preconceptions. An excellent review is by James L. Van Etten, "Giant Viruses," *American Scientist* 99 (2011): 304.

178 *"Megavirus"*: See the discovery paper by D. Arslan et al., "Distant Mimivirus Relative with a Larger Genome Highlights the Fundamental Features of Megaviridae," *PNAS* 108 (2011): 17486–91.

178 *organisms may represent "de-evolved"*: This is not an idle claim, and it is pretty stunning if correct. The research article is by A. Nasir, K. M. Kim, and G. Caetano-Anolles, "Giant Viruses Coexisted with the Cellular Ancestors and Represent a Distinct Supergroup Along with Superkingdoms Archaea, Bacteria and Eukarya," *BMC Evolutionary Biology* 12 (2012): 156.

179 *Paul Davies*: A very nice paper by Davies et al. explores the shadow life idea: "Signatures of a Shadow Biosphere," *Astrobiology* 9 (2009): 241–49. Although these ideas have received a lot of criticism, and I personally think there are some very fundamental problems, it's good stuff to think about.

181 *molecules of arsenate*: Strictly speaking, arsenate is a molecular grouping attached to other things; its chemical formula is $AsO_4{}^{3-}$—an ion. Some organisms *do* incorporate arsenic into organoarsenic molecules—for example, certain marine algae and bacteria. But this behavior appears limited in scope.

181 *from energy transport molecules*: The molecule adenosine triphosphate (ATP, chemical formula $C_{10}H_{16}N_5O_{13}P_3$) is sometimes called the molecular unit of currency for intracellular energy transport. Processes such as photosynthesis or fermentation create ATP, which is then used in numerous other places in a cell, which convert it back to its precursor molecules while extracting energy—in other words, it is a central part of metabolism.

181 *one species of bacterium*: The published name is GFAJ-1, which has been reported to stand for "Give Felisa A Job," in reference to Felisa Wolfe-Simon, who as a postdoctoral researcher was the lead author of the journal article describing this work. Wolfe-Simon et al., "A Bacterium That Can Grow by Using Arsenic Instead of Phosphorus," *Science*

332 (2010): 1163–66. But you cannot read this without seeing the scientific community's response, and there's indeed some potent criticism—a good sanity check comes from B. P. Rosen, A. A. Ajees, and T. R. McDermott, "Life and Death with Arsenic," *BioEssays* 33 (2011): 350–57.

182 *My quoted reaction*: From an interview by Dennis Overbye for *The New York Times* news story, published December 2, 2010, these words were widely requoted.

183 *four-thousand-fold preference*: See M. Elias et al., "The Molecular Basis of Phosphate Discrimination in Arsenate-Rich Environments," *Nature* 491 (2012): 134–37. An earlier paper also finds no evidence for arsenic uptake into viable DNA in the bacterium: M. L. Reaves et al., "Absence of Detectable Arsenate in DNA from Arsenate-Grown GFAJ-1 Cells," *Science* 337 (2012): 470–73.

7: THERE'S SOMETHING ABOUT HERE

185 *Kepler-47*: The discovery of this system was announced in 2012. Both planets in this case are large; one may be a gas giant, the other slightly larger than Neptune. The larger orbits the twin suns every 50 Earth days, the smaller every 303 Earth days. One star is Sun-like, the other is about a third the size.

187 *likely occurs more often*: Indeed, some astronomers have stated that the default type of planetary system in our galaxy is one consisting of several planets in smallish orbits taking days to weeks to go around their stars.

187 *have just such arrangements*: An example is the proposed planetary system of HD 10180, a Sun-size star about 127 light-years from us. Analyses of transit data (yet to be verified with new measurements at the time of writing) suggest that there could be at least nine planets—seven orbiting within 0.5 times the distance of the Earth from the Sun, and two orbiting about 1.5 and 3.5 times farther out. See Tuomi, "Evidence for Nine Planets."

188 *a home* moon: In our solar system the planet Jupiter harbors sixty-seven moons. Most are tiny, but the Galilean moons, Io, Europa, Ganymede, and Callisto, are huge. Ganymede has a larger diameter than the planet Mercury. Astronomers strongly suspect that "exomoons" must exist, and the hunt is on for them. The possibility that they are "habitable" has long been considered; indeed, I wrote about this a while ago, and this paper contains references to several earlier works: C. A. Scharf, "The Potential for Tidally Heated Icy and Temperate Moons around Exoplanets," *The Astrophysical Journal* 648 (2006): 1196–1205.

188 *state of synchronicity*: This is also often termed "tidal lock" or "captured rotation."

192 *rather dryly put it*: Kuhn makes this statement in his book on the Co-
     pernican Revolution, referenced above in the notes to chapter 1. It's also
     well worth looking at the book by Owen Gingerich, *The Book Nobody
     Read*. In this account of his herculean effort to trace the first-edition
     copies of *De revolutionibus*, Gingerich details how Galileo, Kepler, and
     others made use of the book, even annotating it. The book was in fact
     widely read and appreciated by those who could cope with its dense and
     technical nature.

194 *Mars's orbit shifts over time*: See, for example, J. Laskar et al., "Long
     Term Evolution and Chaotic Diffusion of the Insolation Quantities of
     Mars," *Icarus* 170 (2004): 343–64.

195 *patches of interstellar space*: At the moment, our solar system appears to
     be passing through a very tenuous region of material—called, rather
     originally, the Local Interstellar Cloud. It's about 30 light-years across
     and contains about one atom per three cubic centimeters of space. We've
     been in it for somewhere between 40,000 and 150,000 years and will
     probably not emerge for another 20,000 years. Thicker interstellar
     clouds, like the molecular clouds that produce stars, are one hundred to
     one thousand times denser on average.

196 *"rare Earth" hypothesis*: One of the most well-known discussions of this
     idea is in the book by Peter D. Ward and Donald Brownlee, *Rare Earth:
     Why Complex Life Is Uncommon in the Universe* (New York: Coperni-
     cus/Springer-Verlag, 2000). It draws on numerous lines of evidence to
     argue that complex-celled and intelligent life is very unusual in the uni-
     verse; a key piece of the argument is that complex-celled organisms re-
     quire a number of specialized environmental parameters and biological
     components. A more up-to-date, and more astrophysical, take on some
     of these ideas is John Gribbin, *Alone in the Universe: Why Our Planet Is
     Unique* (Hoboken, NJ: John Wiley & Sons, 2011).

197 *thwart the chain of events*: Although I don't discuss this in the main
     text, there seems to be an implicit assumption in rare Earth ideas that
     everything is "perfect" for complex-celled, intelligent life here on Earth.
     I'm not sure it is in the long term. Take, for example, the existence of
     fossil fuels—vast tracts of coal and gas laid down during the Carbonifer-
     ous period around 300 million years ago. These fuels have helped humans
     become what they are today as technological beings. It took very specific
     circumstances of low sea levels, bark-rich trees, and climate change (per-
     haps helped by continental drift and mountain building) to provide this
     energy store. But fossil fuels may have set us up for catastrophic failure
     within the next few centuries. If we're just a blip on the map of evolution,
     I don't think Earth is well-tuned for us—it's simply good enough to allow
     organisms of our ilk to appear momentarily.

199 *Nick Lane and Bill Martin*: Their discussion is in N. Lane and W. Martin, "The Energetics of Genome Complexity," *Nature* 467 (2010): 929–34. Another discussion of the pathway to complex life is by J. A. Cotton and J. O. McInerney, who talk about the "ring of life" rather than the tree of life: "Eukaryotic Genes of Archaebacterial Origin Are More Important Than the More Numerous Eubacterial Genes, Irrespective of Function," *PNAS* 107 (2010): 17252–55.

200 *panspermia—or "all seed"*: This idea goes back a long way, even to the ancient Greeks. It was discussed in the 1800s by scientists including Kelvin and Helmholtz, and in the early 1900s by Svante Arrhenius. It remains a popular, albeit unverified, idea. Certainly within our own solar system the potential seems high for there to have been "exchange" of biological material due to asteroid-impact-driven ejection of planetary surface matter into space and onto other worlds (hence the interest in meteorites of Martian origin). Whether this would result in viable organisms contaminating different worlds is still unclear.

203 *near-miracle of our existence*: I think one can take the baseball analogy further to more accurately match the situation of life on Earth. Suppose that Joe didn't know the total number of balls hit into the crowd that evening—it could have been one or it could have been thousands. He'd still experience the same issues in trying to assess the odds of getting a ball, because it would still seem pretty amazing. The fact is that when it comes to life in the cosmos we're in that same state of ignorance, and it's even trickier because we don't really know the size of the stadium or how many other spectators (habitable worlds) there are.

205 *one of the great discoveries*: Indeed it is, and it resulted in 2011 Nobel prizes for some of the key scientists. They used measurements of the brightness of extremely distant supernovae to evaluate the way in which the universe's expansion is behaving across cosmic time. It was discovered that some 5 billion years ago the universe went from a decelerating expansion (due to the gravitational effect of all mass) to an accelerating expansion. Numerous other indicators existed, and have been found, to corroborate this.

206 *Lawrence Krauss and Robert Scherrer*: Their scientific article was "The Return of a Static Universe and the End of Cosmology," *General Relativity and Gravitation* 39 (2007): 1545–50. Also, a great popular account by the same authors is "The End of Cosmology?," *Scientific American* 298 (March 2008): 46–53.

207 *rate at which stars*: At the time of writing, the most recent estimates of stellar formation across cosmic time come from D. Sobral et al., "A Large Hα Survey at z = 2.23, 1.47, 0.84 and 0.40: The 11 Gyr Evolution of Star-

Forming Galaxies from HiZELS," *Monthly Notices of the Royal Astronomical Society* 428 (2013): 1128–46.

208 *giant black holes*: By sheer (cough) coincidence, I wrote a book about this: C. Scharf, *Gravity's Engines: How Bubble-Blowing Black Holes Rule Galaxies, Stars, and Life in the Cosmos* (New York: Scientific American/Farrar, Straus and Giroux, 2012).

8: (IN)SIGNIFICANCE

211 *stretches in all directions*: This distance is not the light travel time (13.8 billion years) that is often quoted, but is the *co-moving* distance between us and the observable edge of the universe *at the present* cosmological time (equal to the proper distance at this time). This is the actual physical distance, although people still (mistakenly) quote 13.8 billion light-years as the distance.

213 *terrestrial mammals*: Species of mammals, as well as birds, fish, insects, and most macroscopic multicellular life-forms, all seem to follow a similar distribution of physical sizes, skewed about the mean toward smaller sizes—small creatures are more numerous, but do not get smaller than a certain limit. See, for example, M. Buchanan, "Size and Supersize," *Nature Physics* 9 (2013): 129.

215 *yin and yang*: A concept of the interconnectedness and complementarity of opposing or contrary forces or phenomena—light and dark, hot and cold, active and passive, etc.

216 *Michael Storrie-Lombardi*: An astrobiologist and engineer with an MD degree, Mike is known for pioneering work on the use of artificial neural networks in astronomy, and for a wide range of investigations into everything from image compression to stromatolite patterns to the bioinformatics of the origins of life. His research institute's site is www.kinohi.org/.

216 *Stuart Kauffman*: Almost twenty years ago this groundbreaking thinker wrote a great popular account of the nature of complexity and emergent phenomena: S. Kauffman, *At Home in the Universe: The Search for the Laws of Self-Organization and Complexity* (New York: Oxford University Press, 1995).

218 *My own conclusion*: I guess that's the benefit of being an author and a scientist: you get to make informed speculations. I did not, however, know that this would be the exact conclusion when I started writing this book. The gathering of the evidence presented here has been invaluable in arriving at this point.

220 *what's called convergent evolution*: This can be a tricky topic, sometimes skating close to nonscientific theological arguments about "design." That is absolutely not what I mean here, just to be clear. There are

obviously "convergences" in many, many branches of life here on Earth, and logically it is the nature of evolution that variants within species will be selected by their advantages. Thus, within a finite set of physical and chemical environments, and a specific history on Earth, it makes sense that organisms might "re-invent" similar strategies, even if some of those are highly complex. What's not so clear is to what degree we'd see convergence between life on Earth and life on a planet around another star.

221 *Big Ear telescope*: This rather marvelous radio telescope no longer exists; it was demolished in 1998 to open up land for development (for such civilization-saving things as a golf course and houses). But during its active life, from about 1963 to 1997, it not only participated in the SETI but also in more "ordinary" astronomical research, surveying the skies for objects such as radio-loud quasars. Numerous resources and an online memorial can be found at www.bigear.org/

223 *Jerry Ehman himself*: Ehman has written a very nice, and detailed, summary of the Wow signal and his own efforts to figure out what it was. You can read it here: www.bigear.org/wow20th.htm.

223 *Fermi paradox*: A piece of what led Fermi to his comment is that even if interstellar travel is slow, taking thousands of years to go from one star to another, the Milky Way is old enough (at least 10 billion years old) for now-ancient species to have spread out everywhere. Such considerations also come up in discussions of the Drake equation, a combination of numerical factors that the American scientist Frank Drake famously introduced in 1961 to focus discussions on the search for life in the universe. The factors include the fraction of planets capable of supporting life and the length of time that civilizations might broadcast their presence.

225 *ten times more infrared radiation*: This jump in infrared reflectivity and transparency of terrestrial plant life is also known as the "red edge," because it creates a step or jump in the infrared spectrum at wavelengths greater than 700 nanometers.

227 *Roger Penrose*: His popular account of these ideas can be found in R. Penrose, *The Emperor's New Mind: Concerning Computers, Minds, and the Laws of Physics* (Oxford, UK: Oxford University Press, 1989).

229 *beneath Antarctic ice*: A great example is the study of the extraordinary body of water known as Lake Vostok, two and a half miles beneath Antarctica's ice cap and 160 by 30 miles in size. The water in this subsurface lake may have been isolated from the rest of Earth's environment for tens of thousands of years, possibly longer.

230 *all possible realities*: No one knows how many other universes there might be in a multiverse. Some theories of so-called chaotic inflation (to do with the physics of making a universe expand into something larger)

suggest that 10-to-the-power-of-10-to-the-power-of-16 distinct universes are possible. See, for example, A. Linde, V. Vanchurin, "How Many Universes Are in the Multiverse?," *Physical Review D* 81, no. 083525 (2010): 1–11.

231 *words of Carl Sagan*: These are from C. Sagan, *Pale Blue Dot: A Vision of the Human Future in Space* (New York: Random House, 1994).

ACKNOWLEDGMENTS

When I was a small child I lived out in the English countryside, a quiet rural place full of flora, fauna, soil, water, air, and the occasional strange smell. As I grew up there, transforming from meek kid to slightly less meek teenager, one of my secret (and enormously geeky) passions was to try communing with the universe, to be at one with the infinite, to find my place in the enormity of it all. Maybe it fit with a series of adolescent fantasies about unexplored superhero origins, or a mysterious not-yet-revealed past. Maybe I was just strange, or perhaps many children harbor similar ambitions, I still don't know. But on many evenings I'd excuse myself from the supper table and wander outside just as the sky was darkening enough for the stars to begin to appear. I'd walk away from our house and find a private spot somewhere. In the summer it would often be in the midst of a field of rustling wheat where I could sit or lie hidden from every horizon. And then I'd gaze, trying to open my eyes as much as possible, to find that perfect angle where the cosmic night would envelop me, take me over, fill my mind with its infinite emptiness and reveal its untold truths.

As I perched or sprawled on poking bits of vegetation, it slowly dawned on me that as much as the glittering canopy of stars made me feel tiny and insignificant, the inescapable presence of my immediate surroundings also made me feel that I was a vital component of this tapestry. In the cool dampness of the evening air all the pungent mineral flavors and smells of earth and plants would wash across me. And, although the land seemed still, there was the constant rustling of innumerable small creatures—either settling themselves down for the night or foraging through undergrowth and soil. Occasionally, off in the distance would come the lonely wail of some forlorn farm animal, or the hoot of an equally lonesome-sounding owl.

It was soothing yet also an intensely primal and exciting experience, the universe above as much in thrall to this nightly terrestrial routine as it was aloof and uncaring. Of course I knew that the sensations I experienced about this cosmic order to things had to be somewhat illusory. But they were certainly vivid. Surely I, and anyone else Out There, were more than condiments splashed on a complex universe—we had to have some relevance? Or perhaps not, I would force myself to consider; perhaps we were tragic accidents doomed to yearn for significance when we had none at all.

This childhood experience has stayed with me ever since, and has kept that question fresh in my mind. How do we separate out our powerful experiences of the world from our desire to know our place in the universe? The chapters written here are an attempt to tackle some of that puzzle, armed with what I now know, and what many others have thought and discovered.

As I've written this book I've had many conversations. Some have been with my colleagues—other scientists intent on drilling into an endless array of fascinating natural minutiae, and on taking these minutiae and placing them on a cosmic stage. Other conversations, probably the majority, have been with just about anyone who asked me what I was doing. From friends and acquaintances to strangers on planes, on trains, and in the most unexpected places: the sidelines of a soccer game, the middle of a country lane, halfway up a Norwegian mountain, and in the fragrant cheese aisle of a busy supermarket.

It's these latter conversations that have been the most inspiring and interesting. No one, not a single person, has said to me "I have no interest in our place in the cosmos." In fact, the polar opposite: we all have a deep yearning for truth—particularly the kind of rational truth that science aims for, and continues to work away at as it reveals more and more that we don't understand.

For understanding this from the outset I'd like to thank my

wonderful agent Deirdre Mullane of Mullane Literary Associates, and my equally wonderful editor Amanda Moon at Scientific American/Farrar, Straus and Giroux. Their tireless encouragement and hard work have made the writing process so much easier than it would have been otherwise. Thanks too go to publicists extraordinaire Gregory Wazowicz and Stephen Weil, and the editorial team of Christopher Richards, Daniel Gerstle, and Laird Gallagher. A special thanks is due to Annie Gottlieb, whose remarkable copyediting has yet again come to the rescue.

Years and years ago my friend and fellow scientist Michael Storrie-Lombardi planted the seeds of too many ideas in my impressionable head. For that I am immeasurably grateful. I am also grateful for the opportunity to know and interact with so many great scientists who have, over time (often unwittingly), helped me write this book. An incomplete list includes: Frits Paerels, Arlin Crotts, Fernando Camilo, Gene McDonald, Geoff Marcy, Dave Spiegel, Kristen Menou, Ben Oppenheimer, Daniel Savin, Josh Winn, Linda Sohl, Anthony DelGenio, Denton Ebel. Inspiration has also come from talking with many wonderful writers, filmmakers, and science popularizers along the way: Lee Billings, George Musser, John Matson, Dennis Overbye, Marcus Chown, Ross Andersen, Jacob Berkowitz, Bob Krulwich, Dan Clifton. And twice during the course of writing I've had my mind blown by exposure to the incredible gathering that is SciFoo—thanks go to Tim O'Reilly, Larry Page, and Sergey Brin for making that happen.

Deepest thanks also go to friends and family, including Nelson Rivera, Greg Barrett, Helen and Saul Laniado, Windell Williams, Jeff Sklar, and the dearest people in my life, Bonnie, Laila, Amelia, and Marina.

The philosopher Socrates once said, "The unexamined life is not worth living." Admittedly, this is supposed to have been spoken during the trial for impiety that resulted in his execution, but it's still a damn good line. Therefore, I'd like to finally thank you, the reader, for taking time to examine the many and splendid phenomena that make life possible in the universe.

INDEX

abiogenesis, 92, 95

Acta Mathematica (journal), 97, 100, 248

Aldebaran, 256*n*

Allende meteorite, 55, 243*n*

Almagest (Ptolemy), 14, 235*n*

Alpha Centauri system, 89

amino acids, 53, 55, 128, 133, 136, 138–40, 176, 250*n*

AMY1, 151

Andes, 40, 47, 240*n*

Andromeda galaxy, 26, 208

anthropic principle, 29–38, 92, 95, 183, 196, 214–15, 217, 225, 238*n*, 239*n*; fine-tuning parameters in, 30, 32–38, 96, 219, 225, 226

antimatter, 48, 242*n*

Apollo missions, 109, 173, 243*n*

Apollonius of Perga, 14

archaea, 125–27, 146, 178, 181, 220, 250*n*; genetic material of, 135, 175, 196, 198, 252*n*; "mirror" molecules in, 136; molecular engines of, 131; in sea water, 134

Archimedes, 234–35*n*

Arecibo Observatory, 75–77, 94, 245*n*

Aristarchus, 11–14, 16, 186, 234–35*n*

Aristotle, 12–14, 16, 21, 122, 157, 235*n*, 236*n*

Armillaria, 212

Arrhenius, Svante, 260*n*

ASPM, 151

asteroid belt, 59, 60, 114, 240*n*

asteroids, 44–45, 48–50, 64, 115, 120, 173, 197; collision of, 48, 49; fragments of, *see* meteorites; ideas about possibility of life on, 158; impacts with Earth of, 61, 134, 218; transfer of matter across interstellar space from, 229, 269*n*

Astronomia Nova (Kepler), 22

astrophysics, 67, 79, 230, 237*n*; of Big Bang, 205, 214; and life on Earth, 199, 203; planetary, 80, 90, 93; stellar, 137, 209–10

atoms, 53, 82, 128, 129, 137, 152, 216, 228, 237*n*, 238*n*, 248*n*, 265*n*; ancient Greek concept of, 13, 19, 23, 25, 114, 156, 187; atmospheric, 244*n*; in biochemistry, 180; bonding of, *see* molecules; gaseous, 48, 62, 81, 83, 137, 197; of humans, 146; manipulation of, 228; nuclei of,

ALLEN LANE
an imprint of
PENGUIN BOOKS

Recently Published

David Wootton, *The Invention of Science: A New History of the Scientific Revolution*

Christopher Tyerman, *How to Plan a Crusade: Reason and Religious War in the Middle Ages*

Andy Beckett, *Promised You A Miracle: UK 80–82*

Carl Watkins, *Stephen: The Reign of Anarchy*

Anne Curry, *Henry V: From Playboy Prince to Warrior King*

John Gillingham, *William II: The Red King*

Roger Knight, *William IV: A King at Sea*

Douglas Hurd, *Elizabeth II: The Steadfast*

Richard Nisbett, *Mindware: Tools for Smart Thinking*

Jochen Bleicken, *Augustus: The Biography*

Paul Mason, *PostCapitalism: A Guide to Our Future*

Frank Wilczek, *A Beautiful Question: Finding Nature's Deep Design*

Roberto Saviano, *Zero Zero Zero*

Owen Hatherley, *Landscapes of Communism: A History Through Buildings*

César Hidalgo, *Why Information Grows: The Evolution of Order, from Atoms to Economies*

Aziz Ansari and Eric Klinenberg, *Modern Romance: An Investigation*

Sudhir Hazareesingh, *How the French Think: An Affectionate Portrait of an Intellectual People*

Steven D. Levitt and Stephen J. Dubner, *When to Rob a Bank: A Rogue Economist's Guide to the World*

Leonard Mlodinow, *The Upright Thinkers: The Human Journey from Living in Trees to Understanding the Cosmos*

Hans Ulrich Obrist, *Lives of the Artists, Lives of the Architects*

Richard H. Thaler, *Misbehaving: The Making of Behavioural Economics*

Sheldon Solomon, Jeff Greenberg and Tom Pyszczynski, *Worm at the Core: On the Role of Death in Life*

Nathaniel Popper, *Digital Gold: The Untold Story of Bitcoin*

Dominic Lieven, *Towards the Flame: Empire, War and the End of Tsarist Russia*

Noel Malcolm, *Agents of Empire: Knights, Corsairs, Jesuits and Spies in the Sixteenth-Century Mediterranean World*

James Rebanks, *The Shepherd's Life: A Tale of the Lake District*

David Brooks, *The Road to Character*

Joseph Stiglitz, *The Great Divide*

Ken Robinson and Lou Aronica, *Creative Schools: Revolutionizing Education from the Ground Up*

Clotaire Rapaille and Andrés Roemer, *Move UP: Why Some Cultures Advances While Others Don't*

Jonathan Keates, *William III and Mary II: Partners in Revolution*

David Womersley, *James II: The Last Catholic King*

Richard Barber, *Henry II: A Prince Among Princes*

Jane Ridley, *Victoria: Queen, Matriarch, Empress*

John Gray, *The Soul of the Marionette: A Short Enquiry into Human Freedom*

Emily Wilson, *Seneca: A Life*

Michael Barber, *How to Run a Government: So That Citizens Benefit and Taxpayers Don't Go Crazy*

Dana Thomas, *Gods and Kings: The Rise and Fall of Alexander McQueen and John Galliano*